HamSat

Amateur Radio Satellites Explained

By
Pierluigi Poggi, IW4BLG

Radio Society of Great Britain

Published by the Radio Society of Great Britain, 3 Abbey Court, Fraser Road, Priory Business Park, Bedford MK44 3WH. Tel 01234 832700 www.rsgb.org

First published 2015

© Radio Society of Great Britain, 2015. All rights reserved. No part of this publication may be reproduced, stored in a retrieval system, or transmitted, in any form or by any means, electronic, mechanical, photocopying, recording or otherwise, without the prior written permission of the Radio Society of Great Britain.

Edited and Layout: Mike Browne, G3DIH

Cover Design: Kevin Williams, M6CYB

Production: Mark Allgar, M1MPA

Printed in Great Britain by

The opinions expressed in this book are those of the author and are not necessarily those of the Radio Society of Great Britain. Whilst the information presented is believed to be correct, the publishers and their agents cannot accept responsibility for consequences arising from any inaccuracies or omissions.

ISBN: 9781 9101 9307 5

Contents

Foreword v

Introduction 1

1   OSCAR's Story 3

2   The Ground-Space-Ground Propagation 17

3   The Link Budget 27

4   The Ground Station 31

5   Ground Station Antenna 57

6   The Doppler Effect 85

7   Frequency and "Mode" of Amateur Satellites 91

8   The Keplerian Elements 95

     Glossary 104

     Appendix 1 (Associations Dedicated to Amateur Radio Satellites) 112

     Bibliography 115

     Index 119

Acknowledgment

Many People have contributed, with their passion, willingness and expertise, in the drafting and revision of this book.
To all of them my most sincere thanks!

I wish to mention in pure alphabetical order:

- Fabio Azzarello IZ5XRC
- Feng Xu
- Franck Tonna F5SE
- Gaspare Nocera I4NGS
- Graham Shirville G3VZV
- Leif Gustavsson SM6LPG
- Lionel Edwards VE7BQH
- Manuela Mezzetti
- Marco Bruno IK1ODO
- Michele Senestro I1TEX
- Roberto Gerolin IV3CYF
- Sergio Mariotti IK4JGD

To all who have contributed, I extend my most heartfelt thanks.

Pierluigi Poggi, IW4BLG
2015

Foreword

It is now more than fifty years since the first amateur radio satellite, *OSCAR-1*, was placed into orbit.

In the late 1960s, Arthur Gee, G2UK, the then Chair of *AMSAT*, authored a booklet entitled *25 years of amateur spacecraft*. Since that time there has not been a comprehensive record of the activities and achievements of radio amateurs in the field of space communication. This publication will very usefully fill that void.

It clearly demonstrates the wide range of activities that have been undertaken by many groups and individuals around the world.

The invention of the *store and forward* microsats, that used 9k6 GMSK for the first time, enabled many, in the pre-internet days, to communicate with messages and pictures quickly over vast distances.

The three 'Phase B' satellites were truly international spacecraft, and when they reached orbit they seemed miraculous as they enabled long distance communication to take place via transponders that remained in range for more than ten hours at a time. Sadly, although many have tried, it has not yet been possible to find any launch opportunity for follow-on missions.

Perhaps sometime in the future we may have the opportunity to fly an amateur satellite payload to a geostationary orbit. That event would certainly change the face of amateur satellite operation.

Presently, there is a considerable growth in the number of *CubeSats* being launched by many countries - normally by University led groups.

Whilst many of these do not have a direct relationship with amateur radio as they do not include transponders, they do provide tremendous interest to a number of amateurs who receive and decode the telemetry they provide. In addition, many of the members of the university teams find out more about amateur radio as part of their project and go on to obtain their own licences.

The cross-over between educational outreach and amateur radio, at both university and school/college level, gives us the opportunity to help ensure the availability of engineers of many disciplines for the future.

This is surely a vital contribution for us to make.

My congratulations to the editor and contributors of *HamSat*
Graham Shirville, G3VZV

Introduction

Satellite communications all over the globe play a very important role in modern day society and radio amateurs have always been a proud part of that.

It has been over 50 years since the launch of the first amateur satellite, and a lot of the technology that we use in our daily lives, can be attributed to the efforts of these first pioneer experimenters.

The use of technology per se, does not increase the people's cultural experience, but can make the sometimes dry and complex acquisition of knowledge quite the opposite, as has been the experience of radio amateur enthusiasts. A path surely made of ingenuity, but mostly of curiosity to know and explore new frontiers.

What was, in the early years of amateur satellites, a subject area confined to an elite group of enthusiastic engineers with vast experience, has, over the decades, become common ground for all, and especially younger radio amateurs.

This book aims, by offering friendly advice and support, to lead the reader on a wonderful journey of social and cultural growth starting from the basics of amateur satellites.

The complex terms used in conjunction with satellite technology can sometimes scare the reader and become a source of difficulty. In order to minimise an misunderstanding.

Lauch vehicle for Oscar 1

I have tried to present even the most complex topics, in user-friendly way, so as to make them easier to understand.

Having addressed the problems of technical terms, it is time to consider history. Over the years, the successes of the satellite programme has been achieved by vast groups of enthusiasts: technicians, scientists and students, who did not operate in isolation, but gathered together in groups and associations all over the world to form the organisation called AMSAT.

This book will also explainabout the operation modes, their evolution, the frequencies used and their evolution over time, indeed past and future trends, all relevant to understanding what has gone before and where we are heading.

The evolution of technology has enabled unimaginable results and especially the opening of this world to education, from the first degree up to the highest. Today it's the "era of Cubesat", small satellites that fit in the palm of your hand, often designed by students for scientific, educational and communication purposes.

The books also tries to answer questions such as: How do I access an Amateur Satellite? What resources are needed? Is it difficult? How can I know where to beam my antenna?

Hamsat - Amateur Radio Satellites Explained

Over the years, talking with friends who shared my passion and contributing to forums and specific discussion groups have made me aware of the most common questions and lack of knowledge that concern those new to this world. This is of reference for everybody, to read and study, to keep at your side at all times.

I hope the reader will share my fascination for this subject.

May my passion for satellites be with you. Enjoy!

Pierluigi Poggi, IW4BLG
2015

OSCAR's Story

Since the launch of OSCAR I in 1961, it has been a tradition for amateur satellites to carry the name OSCAR standing for "Orbiting Satellite Carrying Amateur Radio".

Upon the request of the group "Project OSCAR", the association AMSAT-NA (NA stands for North America) now manages the numbering of amateur satellites according to the following rules (extracted from the Regulation AMSAT-NA):

1. The use of satellite frequency in the amateur bands must have been coordinated before the launch through an IARU/AMSAT bureau.
2. The satellite must have reached the planned orbit and have been activated successfully.
3. Once in orbit, the satellite must have activated one or more transmitters on the amateur frequency.
4. Once the above requirements are met, the organisation who has built or is the owner of the satellite, may apply for the assignment of an OSCAR serial number through AMSAT-NA, according to a predetermined format.

It is worthwhile to note that it is not necessary to obtain an OSCAR number to legitimise the use of the satellite by the amateur community. The name is still, after 50 years a very proud tradition.

Satellite Phases

Continuous technological development has led the history of amateur radio satellites to be divided into "phases", each of which is characterised by clear objectives and technical or structural characteristics. The following table includes the distinctive characteristics of each "phase".

Phase	Features
1	Satellites powered only by batteries with short life dedicated to technology experiments.
2	Satellites of long duration and capable of translating communications (transponder). LEO orbit powered with solar panels.
3	Satellites of long duration with complex communication systems, control and telemetry. Mostly HEO and Molniya orbits.
4	Satellites in geostationary orbit. These satellites, while appreciated by many and sometimes even designed, have never actually been built and launched.
5	Satellites or even space probes capable of interplanetary or lunar missions. Designed and partially built but never launched.

Although every single satellite is a step forward for science, some have become milestones thanks to the results they have achieved. Let's see now which satellites can be considered the most important in the history of amateur satellites.

1961 OSCAR I

The first amateur satellite, OSCAR I (**Fig. 1.1**), was launched on 12 December 1961, only 4 years after Sputnik, by means of a carrier Thor Agena B from the Vandenberg Air Force Base, California. The orbit was elliptical, 372 x 211km with an inclination of 81.2 degrees and a period of 91.8 minutes.

A group of enthusiasts formed the "Project OSCAR" and convinced the United States Air Force to replace a ballast on the Agena upper stage with the 4.5-kg OSCAR I.

The satellite had a parallel-piped shape and monopole antenna and was powered by batteries. The 140mW transmitter remained in service while the accumulators allowed three weeks during which 570 hams in 28 countries confirmed the receiving of the simple Morse signal "HI-HI" on 144.983MHz. The speed of transmission of the message was controlled by a temperature sensor in the satellite, thus realising a primitive form of telemetry.

OSCAR I re-entered the atmosphere on 31 January, 1962 after 312 orbits.

From 'The Satellite Experimenter's Handbook' [1], Martin Davidoff writes:

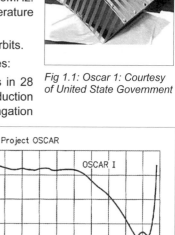

Fig 1.1: Oscar 1: Courtesy of United State Government

"OSCAR I was an overwhelming success. More than 570 amateurs in 28 countries forwarded observations to the Project OSCAR data reduction center. The observations provided important information on radio propagation through the ionosphere, the spacecraft's orbit and thermal satellite design. The OSCAR I mission clearly demonstrated that amateurs are capable of (1) designing and constructing reliable spacecraft, (2) tracking satellites and (3) collecting and processing related scientific and engineering information. Because of its low altitude, OSCAR I only remained in orbit for 22 days before burning up as it re-entered the earth's atmosphere."

1962 OSCAR II

OSCAR II was launched on 2 June 1962, just six months after OSCAR I. Despite the short time since the launch of its predecessor, it was possible to introduce a number of improvements based on experience, among which were: improved thermal insulation to lower the internal temperature, a new system for measuring the internal temperature and a reduction in power to 100MW for extending the battery life.

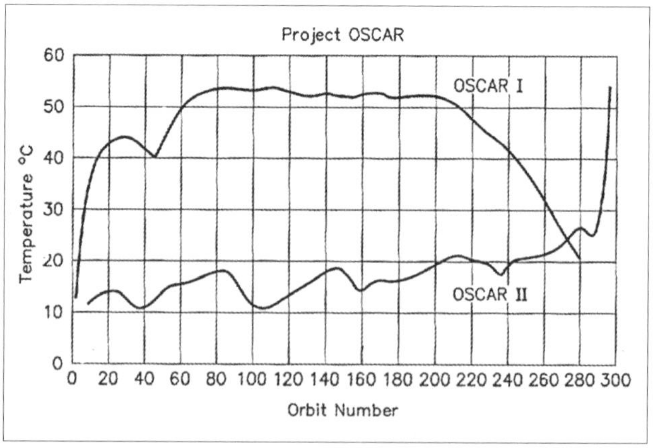

Fig 1.2: Internal temperature comparison between OSCAR I and II. Source: Martin Davidoff, K2UBC The Radio Amateur's Satellite Handbook [2]

The image (**Fig 1.2**) shows the temperature variation of OSCAR I and II over the orbits. It is clear that OSCAR II is significantly colder than its predecessor. The surge in the final temperature is due to its return to the atmosphere.

1965 OSCAR III

OSCAR III was launched on 9 March 1965 using a vector Thor Agena B from Vandenberg Air Force Base, California. Its orbit was elliptical, 924 x 891km with an inclination of 70.1 degrees, while the period of 102.7 minutes with a weight of 16.3kg.

OSCAR III was the first amateur satellite to be powered by solar panels and to retransmit signals received from the ground. OSCAR III was, in fact, the first amateur satellite to retransmit voice communications on the 2m band by means of a linear transponder of 50kHz bandwidth (146MHz up-link and down-link 144MHz). The repeater function of OSCAR III remained in service for 18 months, during which time more than 1000 radio amateurs in 22 countries communicated by means of the transponder. The two beacons continued their function for many years.

1965 OSCAR IV

OSCAR IV was launched on 21 December 1965 by a Titan 3C carrier from Cape Canaveral, Florida. The carrier had a partial failure at the beginning of the climb and was not able to deliver the satellite to the correct orbit. The final orbit of 29120 x 168km was inclined at 26.8 degrees, with a period of 587.5 minutes. It had a weight of 18.1kg and 4 monopole antennas.

OSCAR IV was the first satellite to allow amateur radio links between the Soviet Union and the US, who were still engaged in the Cold War at this time.

OSCAR IV was built by TRW Radio Club of Redondo Beach, California. It had a 3W linear transponder with 10kHz of bandwidth (144MHz uplink and 432MHz downlink). It remained in service until 16 March 1966 and returned to the atmosphere on 12 April 1976.

1970 Australia-OSCAR 5

Australia-OSCAR 5 was launched on 23 January 1970 by a Thor Delta vector from the Vandenberg Air Force Base, Lompoc, California. Its orbit was 1476 x 1431km with an inclination of 101.8 degrees while its period was 115 minutes and its mass 17.7kg (almost 9kg were due to batteries). The antennas were a monopole for the 2m band and a dipole for the 10m.

It was the first amateur satellite with remote control functions on board.

Built by the students of the University of Melbourne, Australia, it was battery powered and transmitted its telemetry on 2m (144.050MHz and 50mW) and 10m (29.450MHz and 250mW).

1972 AMSAT-OSCAR 6

AMSAT-OSCAR 6 was launched on 15 October 1972. It weighed 16kg and the orbit, inclined at 101.7 degrees, was 1450 x 1459km. It had three antennas, a quarter-wave monopole for 144 and 435MHz and a half-wave dipole for 29MHz.

It was the first satellite to carry complex control systems on board to enable communications between satellites (with Oscar 7) and to be used for medical and security services from remote locations around the world.

It had a Mode A (100kHz and 1W) transponder which allowed traffic "store-and-forward" on CW and RTTY (Codestore). Its mission lasted until 21 June 1977 when it experienced battery failure.

1974 AMSAT-OSCAR 7 (Phase-IIB)

It was released on 15 November 1974 by a Delta 2310 from Vandenberg Air Force Base in California.

After 7 years of smooth operation, a battery failure made it out of service. Surprisingly, at 17:28 UTC on 21 June 2002, Pat Gowen, G3IOR, received its signals again! A second fault, probably the power supply, had isolated the batteries and the satellite powered only by solar panels was back in operation! Today it is the oldest amateur satellite on duty, even if a bit sluggish and of unpredictable performance.

1978 AMSAT OSCAR 8, (also known as Phase 2D) built by radio amateurs in the U.S., Canada, Germany and Japan. The primary mission was to provide an educational tool and amateur communications. Launched on the 5 March 1978 it consisted of a 145.90-146.00/435.1MHz (inverted) and 145.85-90/29.4-5MHz transponder using a circularly polarized VHF/UHF canted turnstile antenna system. Telemetry beacons operated on 435.095 MHz and 29.402 MHz using an HF dipole antenna. The satellite operated until the battery finally failed in the middle of 1983

1978 Radio Sputnik RS-1 and RS-2

The first two Russian satellites were launched by a Soviet amateur on 26 October 1978 from the Plesetsk Base with a C1 vector Cosmos 1045. The weight was 40kg and the elliptical orbit was 689 x 1709km with an inclination of 82.55 degrees. They had an inverted V antenna for VHF and a quarter-wave whip for HF. The satellites were equipped with a linear transponder, mode A (up 145MHz, 29MHz down), and a telemetry system similar to the Codestore of AMSAT-OSCAR 6.

1981 UoSAT-OSCAR 9 (UoSAT 1)

UoSAT OSCAR-9 was launched on 6 October 1981 by the carrier Thor Delta. With a weight of 52kg, it was placed on a sun-synchronous LEO orbit of 538 x 541km with an inclination of 97.46 degrees.

It was the first satellite with a computer that managed and monitored all the functions (IHU - Integrated Housekeeping Unit), such as batteries, orientation, remote control and the experiments in S-band. Built by the University of Surrey (UK), it was a satellite for scientific and educational purpose: in fact it had many experiments and beacons but no transponder for amateur radio use. UO-9 remained in service until its destruction re-entering the atmosphere on 13 October 1989. Its S-band beacon represented a novelty and together with the one on X-band was widely described in The Radio and Electronic Engineer, 52, No 8/9, p. 412-416 August / September 1982.

1983 AMSAT-OSCAR 10 (Phase 3B)

AMSAT-OSCAR 10 was launched on 16 June 1983. It was the first amateur satellite designed to reach the final orbit by means of autonomous propulsion systems. Intended for a Molniya orbit type, due to a collision with some parts of the carrier at the time of detachment, it entered instead into a low inclination orbit. In December 1986, the main computer was irreparably damaged because of radiation causing the loss of satellite control and in particular its orientation. It remained randomly in operation according to sun exposure and the "wake up" CPU mode, in a similar manner to what happens today to OSCAR 7.

Radio Sputnik RS-3, RS-4, RS-5, RS-6, RS-7 and RS-8

These six Russian satellites were placed in orbit at the same time from a single carrier on 17 December 1981. RS-3 and RS-4 were experimental satellites and had no transponder while the others were equipped with radio gear for the A-mode.

RS-5 and RS-7 possessed a particular device named ROBOTS which allowed the connection (QSO) with the satellite. A typical connection with the ROBOT began with a call from the ground station on the up-link frequency to which the satellite responded on the down-link frequency with a brief message in CW containing the sequential number of the QSO.

1986 Fuji-OSCAR 12 (JAS-1a, Fuji)

Fuji-OSCAR 12 was launched on 12 August 1986 with the first flight test of the vector H-1. Its weight was 50kg with a circular orbit of 1479 x 1497km inclined at 50 degrees.

FO-12 was the first Japanese amateur satellite developed by JARL (Japan Amateur Radio League). The on-board system development and the subsequent integration were carried out in the laboratories of NEC. FO-12 remained in service until 5 November 1989 when a battery failure made it unusable.

1988 AMSAT-OSCAR 13 (Phase 3C)

AMSAT-OSCAR 13 was launched on 15 June 1988 with the first test flight of the new Ariane 4. With a weight 92kg plus 50kg of fuel, its orbit was Molniya with an inclination of 57.4 degrees.

AO-13 was the third satellite of phase 3 and was named for this "Phase-3C". Built by an international team of amateurs led by Dr. Karl Meinzer DJ4ZC, (AMSAT-Germany), it was equipped with four beacons, one digital and four linear transponders called Rudak-1.

It remained in service until its return into the atmosphere on 5 December 1996.

1997 Radio Sputnik 17

RS-17 was a scale model of the very first artificial satellite built by some students to commemorate the 40th anniversary of the launch of Sputnik I.

It was put into orbit manually by a Russian cosmonaut from the MIR space station. RS-17 broadcast its beep-beep for 55 days with the last reception taking place around 21.00UTC on 29 December 1997.

2000 AMSAT-OSCAR 40 (P3D)

AMSAT-OSCAR 40 was launched on 16 November 2000 by an Ariane 5 rocket. With an orbit of 1157 x 58665km and an inclination of 24.7 degrees, probably it was the most bulky amateur satellite ever built and launched. Its 244kg weight can put its complexity in evidence. It was unfortunately the last great amateur satellite to be launched. A joint project of AMSAT-DL and AMSAT-NA, it has been the largest and most complex, comprehensive and innovative amateur satellite ever put into orbit. Shortly after the launch, a valve failure created initial damage to the structure and the antennas limiting its operational capability. In January 2004 it came to a sad end: a main battery failure caused the failure of other vital on-board systems making it permanently inoperable.

2004- AO-51

It was launched on 29 June 2004 in Baikonur Cosmodrome, Kazakhstan, on a Dnepr launch vehicle. It has a sun synchronous low earth orbit of 823 x703km with an inclination of 98.1 degrees and an orbital period of 100 minutes. This satellite, formerly known as ECHO and built by AMSAT-NA, was equipped with an FM repeater with 144MHz and 1.2GHz uplinks as well as 435MHz and 2.4GHz downlinks. It also contained a digital subsystem that transmitted telemetry and provided a complete PACSAT BBS. As well, there was a 10m PSK uplink. The AO-51 satellite was easily workable with an amateur VHF dual band hand-held radio, which led to its increased popularity. As of May 2011, the satellite faced battery problems which resulted in irrecoverable failure on 29 November 2011.

2005-VO52 HAMSAT

It was launched by PSLV-C6 on 5 May 2005 and placed into a polar sun synchronous orbit of 97.89 degrees of inclination. This satellite is known also as HAMSAT or HAMSAT INDIA and VU2SAT and it is a micro-satellite weighing 42.5kg. Its peculiarity is to carry two independent transponders, one built by William Leijenaar (Call Sign: PE1RAH), and the other by enthusiasts at ISRO (Indian Space Research Organisation).

2009-HO68 Hope OSCAR

On 15 December 2009, AMSAT China launched a new amateur radio satellite, recently designated as Hope Oscar (HO) 68 but formerly known as the "Nozomi" Olympic Star Code XW-1. It is on a sun-synchronous orbit of about 1200km, inclined 105 degrees. This satellite not only has a V/U FM repeater but also a V/U linear transponder for SSB and CW as well as a packet BBS system. It is the most popular amateur satellite so far built by China.

2013-AO73 FunCube

Successfully launched from Russia on a DNEPR rocket on Nov 21st 2013 at 07:11:29 UTC is a complete educational single CubeSat project with the goal of enthusing and educating young people about radio, space, physics and electronics. FUNcube-1, now registered as a Dutch spacecraft, is performing very well. More than 500 stations around the world are already receiving and decoding the telemetry and many schools are already involved.

Let's See Some Statistics of the Evolution of Amateur Satellites.

Frequency

The first aspect is the trend of the frequency used for uplink and downlink on the satellites launched every year, (**Fig 1.3**).

As you can see, today the most predominant ones are the pair of V-U 144 & 435MHz. While HF has been almost abandoned, microwaves are still used occasionally with more and more interest in the recent smaller CubeSat.

It is worth noting that over the years more and more amateur satellites have been launched thanks to technological progress and space knowledge promotion all over the world. In the last decade we have reached the level of almost 8 amateur satellites launching for each year! (**Fig 1.4**)

During the recent years of deep global economic downturn, radio amateurs and students have never stopped working on how to improve satellite systems and have found out satellite downsizing is the way to overcome a lack of economic resources. A new generation of satellites has thus been born: the CubeSat!

Some weight and size considerations that must be observed when developing satellites for amateur radio use, are as follows.

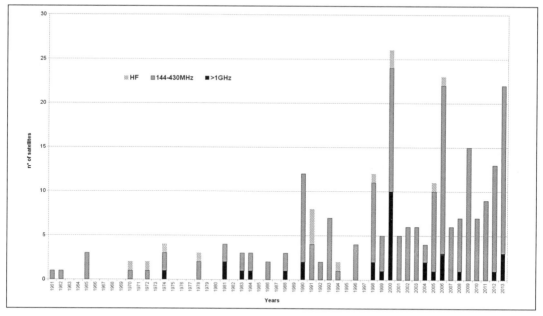

Fig 1.3: Trend of frequency bands used for space communications

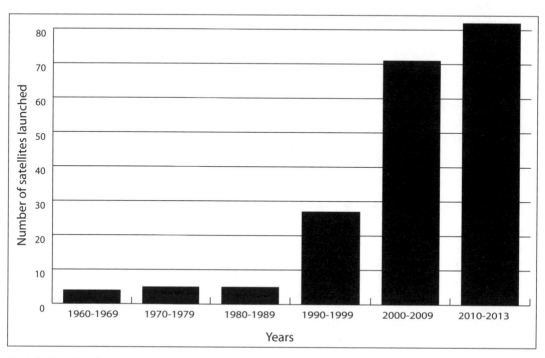

Fig 1.4: Rising trend of amateur satellites being launched

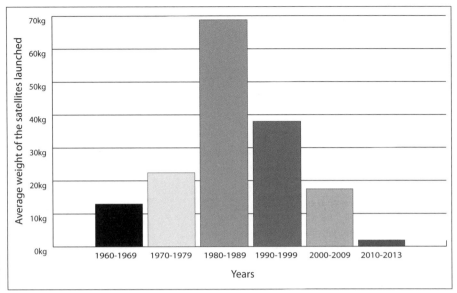

Fig 1.5: Trend of the amateur satellites' weight

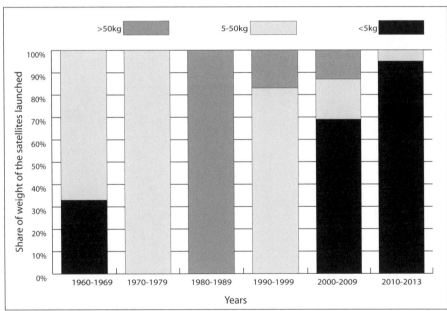

Fig 1.6: Trend of the share of weight class of amateur satellites

Weight

The analysis shows that, after an initial phase (culminating in the 1980s) of continuous increase in weight of satellites (see **Fig 1.5 & 1.6**) which went hand in hand with an increase of their complexity, there has recently been a trend towards weight reduction due to higher integration and budget constraints.

Nowadays, the most popular satellite has a weight of less than 5kg.

Size

This analysis confirms a similar trend to that which has already been witnessed in terms of weight. During the last 10 years, amateur satellite size has reduced enormously, partially due to the development and popularity of CubeSat. (see **Fig 1.7 & 1.8**)

Which orbit?

Let's have a look now at the kind of orbit typically used by amateur satellites with a special focus on the apogee/perigee parameters:

Every "dot" in the graph is a satellite. As you can see (**Fig 1.9**), many lie on, or near the diagonal line which represents the circular orbit. The major number of the dots is in the grey area related to the LEO orbit while very few are in the MEO or even HEO (light grey or white background).

Another interesting parameter of the orbit to be analysed is the eccentricity, which over the years has changed as follows in **Fig 1.10**.

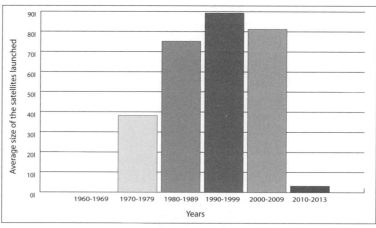

Fig 1.7: Trend of amateur satellites size

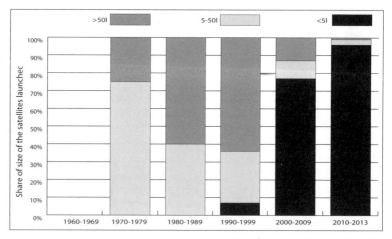

Fig 1.8: Trend of the share of amateur satellites size

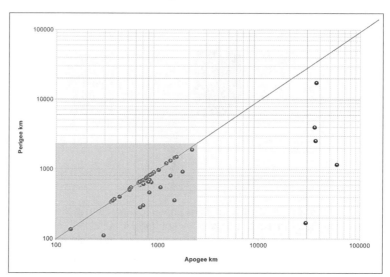

Fig 1.9: Apogee and perigee pairing of the most known amateur satellites

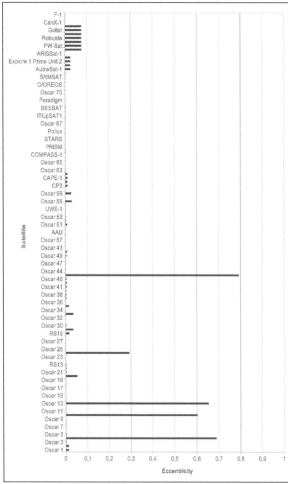

Fig 1.10: Eccentricity of the orbit of different satellites including amateur satellites

While in a statistical form it can be displayed as **Fig 1.11**.

From the graph (Fig 1.11) you can see the most popular is the circular one with the remaining 25% shared by low and highly elliptical.

Lifetime

How long can an amateur satellite be in service? This is a good question and is quite challenging to answer. A satellite cannot be repaired in the harsh environment of space. Let's see now how long the lifetime has been of some satellites **Fig 1.12**.

As you will notice, unfortunately several satellites have had a very short lives, but there have also been many cases of satellites remaining in service for several years, without any trouble.

These are the performances of the satellites which are already "retired". Then what about those which currently still are working?

Let's see how those active satellites are behaving as of 1 April 2014 (**Fig 1.13**)

Please do not be fooled by the small figures for some satellites as they are still new and are yet to accumulate long working periods. The general tendency is quite good with many satellites able to last for a long period of continuous service, including many Cube-Sats.

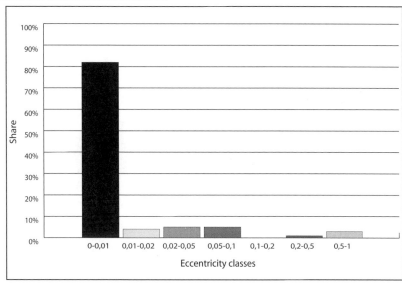

Fig 1.11: Graph showing the eccentricity of the orbit of amateur and commercial satellites

Hamsat - Amateur Radio Satellites Explained

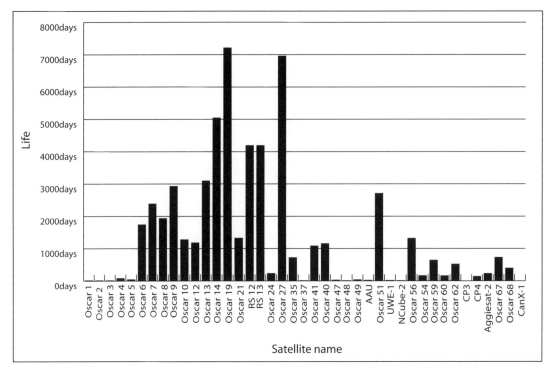

Fig 1.12: Lifetime of the most popular satellites in the past

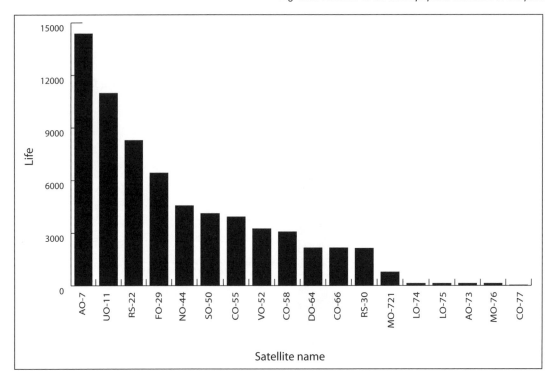

Fig 1.13: Lifetime of the satellites in service as of 1 April 2014

Oscar's Story

Which amateur satellites are still active today?

This question is hard to answer because of the high rate of list updating. Almost every month, some new satellites become active while some others, unfortunately, cease to operate.

Let's check the list as of 31 March 2014 (credits: DK3WN web site), taking into account those active and partially or occasionally active ones:

Satellite	NORAD	Uplink	Downlink	Beacon
AO-7 (Phase-2B	07530	145.850-950	29.400-500	29.502
AO-7 (Phase-2B)	07530	432.125-175	145.975-925	145.970
UO-11 (UoSAT-2)	14781	.	145.826	
RS-22 (Mozhayets)	27939	.	.	435.352
FO-29 (JAS-2)	24278	145.900-999	435.900-800	435.7964
ISS	25544	437.550	437.550	437.550
ISS	25544	145.20/144.49	145.800	
ISS	25544	145.825	145.825	145.825
NO-44 (PCsat1)	26931	145.827	145.827	145.827
SO-50 (SaudiSat-1c)	27607	145.850	436.795	.
VO-52 (Hamsat)	28650	435.225-275	145.925-875	145.860
CO-55 (CUTE-I)	27844	.	437.470	436.8375
CO-57 (XI-IV)	27848	.	437.490	436.8475
CO-58 (XI-V)	28895	.	437.345	437.465
DO-64 (DELFI-C3)	32789	.	145.870	145.8675
CO-66 (SEEDS-II)	32791	.	437.485	437.485
RS-30 (Yubileiniy)	32953	.	435.315/215	435.315
MaSat-1 (MO-72)	38081	.	437.345	437.345
AO-73 (FUNCube-1)	39444	435.150 - 435.130	145.950 - 145.970	145.935
LO-74 (CubeBug-2)	39440	.	.	437.445
MO-76 ($50SAT,Eagle-2)	39436	.	.	437.505
LO-75 (CAPE-2)	39382	145.825	145.825	145.825
CO-77 (ARTSAT-1 INVADER)	39577	.	437.200	437.325
SO-67		145.875	435.345	435.300
SwissCube-1			437.505	437.505
BeeSat			436.000	436.000
ITU-pSat1			437.325	437.325
HO-68		145.925-975	435.765-715	435.790
HO-68		145.825	435.675	435.790
Tlsat-1		145.980	437.305	145.980
O/OREOS			437.305	437.3037
FO-70		435.025	145.825	145.825
FO-70		437.345	145.825	145.825
Jugnu			437.505	437.2759
SRMSAT		145.900	437.500	437.425

Satellite	NORAD	Uplink	Downlink	Beacon
RAX-2		437.345	437.345	
AO-71		437.475	437.473	
E1P-U2		437.505	437.502	
M-Cubed		437.485	437.485	
MaSat-1		437.345	437.345	
Xatcobeo		437.365/145.940	437.365	
PW-Sat1	435.020	145.900	145.902	
Horyu-2		437.375	437.378/372	
PROITERES		437.485	437.485	
Aeneas		437.600	437.600	
CSSWE		437.349	437.349	
CP5		437.405	437.405	
FITSAT-1		437.445/5.8GHz	437.250	
TechEdSat		437.465	437.465	
WE_WISH		437.515	437.505	

As a conclusion of this chapter, it is worth exploring briefly the sites and orbital vectors which, over the years, have "made history", especially with regard to amateur satellites.

The Main Launch Bases
Many launch bases have been used for amateur satellite launching - as shown briefly below:

Baikonur
Is the largest former Soviet Union cosmodrome and the only one used for astronauts launching. After the collapse of the former Soviet Union, this site has gone through geographical transfer to the state of Kazakhstan while maintaining the official designations NIIP-5 and GIK-5

Cape Canaveral
Is the largest US launch centre used for all the manned missions. Currently, only six of the 40 launch facilities are still operational. In the vicinity of Cape Canaveral there is the Kennedy Space Center in Merritt Island used by NASA for the launch of the Saturn V and the Space Shuttle.

Kodiak
Is the new launch base of the Alaska Aerospace Development Corporation located about 400km south of the Alaska State's capital city, Anchorage, a great location to launch satellites into polar orbit.

Kourou
After granting independent status to Algeria, France tried to look for a new launch site. After lengthy assessments, the choice fell on its South American colony of Guiana. Being a remote island makes it suitable for the launch at almost every angle from -100.5 to 1.5 degrees. Being close to the equator also gives also maximum efficiency for the launch into equatorial orbit (GEO typically).

Plesetsk
Plesetsk was the northern centre of the former Soviet Union, for the launch of satellites mostly for military use. With the collapse of the USSR and the consequent loss of territorial control of Baikonur (Kazakh territory today), Plesetsk could soon become the new Russian missile test centre. Its high latitude will have negative impact for geosynchronous orbits. To overcome this limitation, technicians are considering orbits involving the interference with the moon's gravity to move the vector in the equatorial orbital plane spending the minimum energy.

Sriharikota
It is the main Indian launch site located in the west coast of the Bay of Bengal which has been active since 1971.

Taiyuan
China's launch site for polar-orbiting satellites, also known as Wuzhai Taiyuan Satellite Launch Center (CGT), is situated in Kelan County in the northwest of Shanxi Province, 280km from the city of Taiyuan from which it takes the name.

Tanegashima
It is the main launch site for Japanese carriers N and H class. Active from 1975 to 2007, it has witnessed more than 140 successful launches.

Vandenberg
It is a launch base owned by the US Air Force located in the State of California about 240km northwest of Los Angeles, used for both military and commercial missions to polar orbits without astronauts.

The Main Orbital Carriers (rockets)

Ariane
The first commercial French and European carrier, this project began in July 1973 and ended with the launch of the Ariane 1 rocket eight years later.

Ariane 4
This French carrier was created through continuous improvement projects of Ariane 2 and 3 with the goal of 90% payload capacity increase. This project started in 1982 with a total cost of 476 ECU .

Ariane 5
Although it still belongs to the family of Ariane, Ariane 5 was a completely new design which differs significantly from its predecessors. This project began in 1984 with the first prototype being built in 1988. Initially designed as a vector for the human missions Hermes, it was used for commercial purposes in 2000 after the abandonment of this project with the capacity to carry two satellites in geosynchronous orbit at the same time.

CZ
The first Chinese carrier was the DF-5 (1971), a two-stage rocket similar to the American Titan, the Russian R-36 or the European Ariane. The DF-5 was the "father" of a large family of vectors called 'Chang Zheng' CZ-2, CZ-3 and CZ-4 capable of putting into orbit payloads of up to 9200kg.

Delta
Delta has been the longest-serving, while being reliable and cost-effective, carrier in US history. The initial cost was controlled by redeploying some parts of the Apollo program. This project underwent continuous improvements, increasing its payload capacity from 68kg in 1962 to 3810kg in 2002 on geosynchronous orbits! Thanks to its unquestionable quality, the Delta project survived innumerable attempts to delete or replace with other carriers. Over 1000 Delta had been successfully used as of 2008 after almost 50 years of service.

H-1
A Japanese carrier built upon a Delta licence with specific modifications to the upper stages.

PSLV
Is India's third-generation vector used to launch payloads into polar-orbiting.

Soyuz

A Soviet Union product with more than 1700 launches and an incomparable success rate of 97.5%, it has been the most long-lived, reliable while broadly used, launching vector in history. Since 2000 new versions for business customers have been introduced while the Kourou base has been equipped for its launch since 2009.

Titan

It is a US vector developed by the United States Air Force in the 1960s. Designed to be a cheap vector, it was not able to keep its promise and was soon replaced by the Atlas V and Delta IV system in the US, and by the Ariane project in Europe.

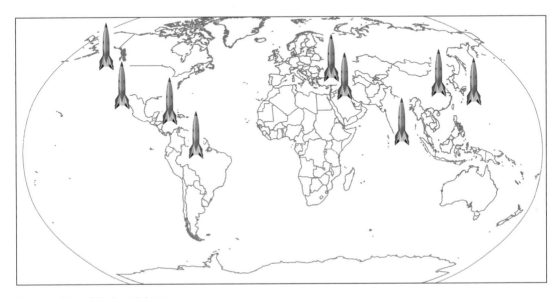

Fig 1.14: Map of the launch bases

Reference

[1] 'The Satellite Experimenter's Handbook' , Martin Davidoff

The Ground-Space-Ground Propagation

Among radio amateurs, we use the term "propagation" to refer to any phenomenon that extends the range of radio communication beyond the optical range.

More generally, and especially in space communications, we refer to all those events that disturb, positively or negatively, the consistency of the signal received.

The radio wave, moving from the satellites down to our stations, must pass through the atmosphere and the earth's ionosphere and is subject to various interactions with those layers.

The thickness of air and ionosphere to be crossed is a function of both the orbital height of the satellite, as well as its position with respect to our station, with a minimum when passing through the zenith (elevation 90 degrees, vertically above us) and a maximum when just rising (or setting) from the horizon.

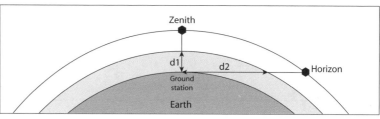

Fig 2.1: Change of the signal path upon changing of the satellite elevation

For a generic angle of elevation of the satellite, the formula becomes:

$$d = \frac{R+Q}{sen(90+\theta)} \cdot sen\left(90 - \theta - arcsen\left(\frac{R \cdot sen(90+\theta)}{R+Q}\right)\right)$$

Where:

θ = angle of elevation in degrees
R = mean radius of the earth (ie 6378km)
Q = orbital altitude of the satellite

Let's take a practical example about a LEO satellite running on an almost circular 350km altitude orbit (the ISS for example). We may consider for simplicity that the troposphere ends at 10km of altitude while the ionosphere would do at 1000km.

Doing the calculations as indicated above, we can have the following:

	Horizon	Zenith
Path in the troposphere [km]	357	10
Total distance [km]	2142	350

Hamsat - Amateur Radio Satellites Explained

We can show the results as follow:

Elevation	Tropo Path	Total
0	357	2142
5	105	1657
10	56	1304
15	38	1053
20	29	876
25	24	747
30	20	652
35	17	581
40	16	526
45	14	483
50	13	449
55	12	422
60	12	401
65	11	384
70	11	371
75	10	362
80	10	355
85	10	351
90	10	350

It can also be shown as the following graphic (**Fig 2.2**).

It is clear that, with the satellite low on the horizon, the situation becomes complicated quickly and the simulation shows the path up to 30 times more dependent on atmospheric conditions than in the case of satellite on our vertical.

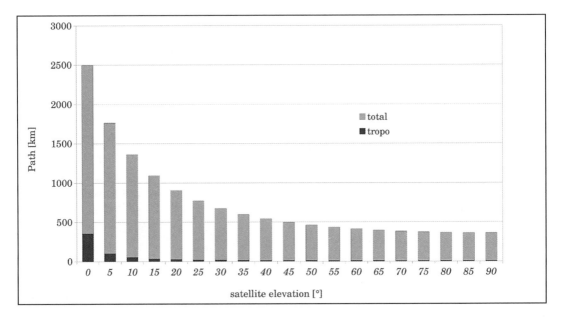

Fig 2.2

The Ground-Space-Ground Propagation

It should be noted also that not only the apparent thickness of a layer affects the radio link but also the angle at which it is approached. The effect is particularly noticeable for example in ionospheres' reflections on shortwave.

We will have to keep in mind some of these aspects when calculating a "real case".

In the case of terrestrial radio traffic, many of the phenomena that we are about to discuss are expected and appreciated by radio amateurs for their properties to extend the range of the contacts compared to average conditions, while in the case of earth-space communications, they do not always play in our favour and may even prevent contact with the satellite.

The layers of the atmosphere

The atmosphere that surrounds our planet is divided, in several ways, each one characterised by the homogeneity of specific properties.

Below are the two subdivisions which are most relevant to our study.

Depending on the temperature and the types of gas (**Fig 2.3**), those who study the atmosphere as a layer of gas around our planet have divided it into two bands:

- the homosphere which lies between the ground and about 100km of altitude where the average composition of the atmosphere does not change, thanks to its continuous motions of vertical mixing
- the heterosphere which starts from 100km of altitude where the condition of diffusive equilibrium prevails influencing how the composition varies with the altitude with increasing presence of lighter gases such as helium and hydrogen.

One of the methods used to divide the earth's atmosphere is based on varying the proportion of its main parameters and first of all the temperature.

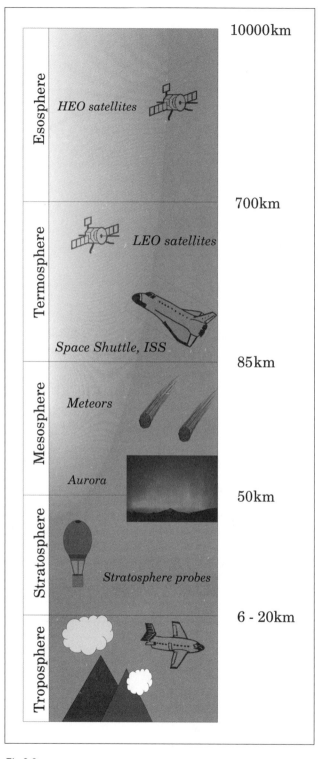

Fig 2.3

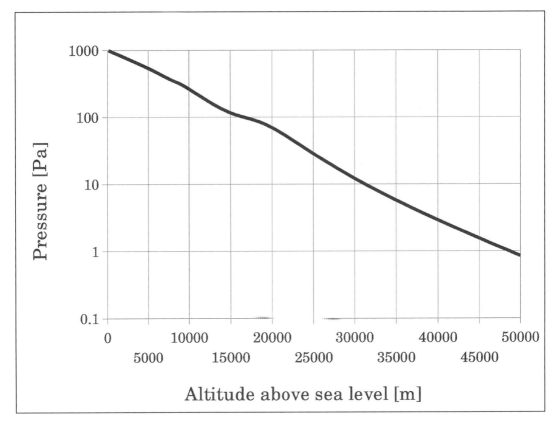

Fig 2.4: Trend of the atmospheric pressure vs the altitude

Troposphere

It is the layer in which we live and in which occur almost all the weather phenomena. It contains 80% of the total gaseous mass and 99% of the water vapour with an average temperature of 15°C at sea level, which decreases with altitude (on average -0.65°C per 100m above sea level) down to about -60°C in tropopause. The troposphere has a thickness that varies according to the latitude: at both of the poles it is approximately 8km thick and about 17km at the equator, a phenomenon caused by the earth's rotation. The atmospheric pressure decreases with altitude. The average data issued by NASA indicates the following values:

Altitude [m]	Pressure [hPa]
0	1013
1000	897.52
2000	795.21
4000	615.9
6000	471.05
8000	354.55
10000	263.38

Altitude [m]	Pressure [hPa]
15000	116.5
20000	69.9
30000	12.16
48500	1.01
69400	0.1

Continuing the ascent, the temperature keeps stable at about -60°C in the tropopause, the transition zone between the troposphere and the stratosphere.

The term tropopause derives from the Greek "τροπος" which means way, mutation and change. We use this term because, in this first layer of the atmosphere, we find all those movements and blending of the air that characterise meteorology. The troposphere is also the place of life: all life forms which are developed in it use some of the gas that is present.

Stratosphere

It is the atmospheric layer which extends above the troposphere up to an altitude of 50-60km. It is characterised by an important phenomenon known as thermal inversion. While in the troposphere the temperature decreases with height in the stratosphere it increases by rising to the temperature of 0°C. This particular phenomenon is due to the presence of a layer of ozone that absorbs most of the solar UV radiation (about 99%) allowing the development of life on our planet. In the stratosphere the components are increasingly rarefied gas, water vapour and dust.

Mesosphere

In this third band, ranging from 50 to 80km in altitude, the atmosphere is no longer suffering the influence of the earth's rotation and it is constant at all latitudes. It is characterised by an increment of light elements at the expense of the heavier ones: carbon dioxide disappears rapidly together with water vapour and also the partial percentage of oxygen begins to decrease with altitude. On the contrary, it increases the fractions of light gases such as helium and hydrogen. In this layer, where the heating effect of the ozone layer is gone, the temperature decreases and reaches the minimum value, variable between -70 and -90°C, at around 80km, an altitude in which it may create the sometimes so-called "noctilucent clouds", probably made of ice crystals and dust: they are visible during the summer, at dusk and appear as thin and bright clouds.

In this layer originate the phenomenon of shooting stars due to the ablation of small meteorites which disintegrate before reaching the ground as the earth crosses their orbits. Beyond the mesopause, at an altitude of about 100km, the atmosphere is too thin to have any further significant resistance anymore to the motion and thus it is possible to position orbiting bodies such as satellites and manned research stations there. For this reason, in astronautics the mesosphere is considered the boundary with space.

Thermosphere

The Thermosphere is the zone in which temperatures will rise, up to values close to 1000°C. Do not be misled by this particularly high figure, atoms and molecules are very much energised and therefore express high temperatures, but their density, which is very low, actually creates a cold environment, close to that of interplanetary space! Physicists, to correctly describe this situation, have introduced the concept of kinetic temperature which applies when the physical conditions of the matter precludes the use of classical macroscopic-operational definition of temperature. For kinetic temperature we mean the temperature at which the gas particles should be on the earth's surface in order to move with a kinetic energy equal to that which they have in the conditions in which they actually are found, directly following the definition of microscopic temperature given by the kinetic theory of gases. It's the layer where many satellites and space stations have their orbits.

Exosphere

This is the outermost part of the atmosphere where only the lighter fractions of gas such as hydrogen and helium are present, often in atomic form and coming from solar jet. The kinetic temperature of the exosphere increases with altitude reaching and sometimes exceeding 2000°C.

Depending on the ionisation

The ionosphere is the layer of atmosphere in which the atmospheric gases are strongly ionised. It contains only a small fraction of the atmospheric gaseous mass, about 1%, but has a thickness of some hundreds of km which luckily absorbs a large part of the ionising radiation coming from outer space.

It has a divided layer structure during the daytime caused by strong solar radiation that ionises the gases differently at different heights, while at night, some of these layers merge together. Radio waves suffer different effects depending upon the frequency, the layer, its ionisation conditions and the angle of impact.

The D layer

This extends approximately from 50 to 90km in altitude with a concentration of electrons which grows rapidly with the height. The concentration of free electrons in this layer presents a major trend which reaches its maximum shortly after local solar noon while showing very low values during night time.

Even during the winter time, despite the high zenith distance at our latitudes from the sun (50-70 degrees), we can often observe very high electronic concentration, mainly between 70 and 90km, probably due to the nature and concentration of the gases on that region.

The D layer can reach a density up to 10 billion electrons per cubic meter at altitudes between 50 and 90km, with a high density of neutral particles too.

This layer does not have large reflective properties for radio waves used in a radio link via ionosphere because of the relatively low electron density, while it assumes considerable importance with regard to their absorption.

So the D layer can be considered as an excellent absorbent layer.

The E layer

Between 90 and 130km is located the layer E, very regular and positioned at a height in which the temperature has an excursion from -80 to +80°C. The electron density depends directly on the zenith distance from the sun. Thus, there is a daily maximum around noon and a season-high in the summer. The maximum concentration of free electrons is placed around 110km and is about 100 billion per cubic metre.

Sporadic E is an important part of the E layer located at about an altitude of 120km, so called because its presence is random and sporadic. It seems that its ionisation is due to meteorites and cosmic phenomena not related to solar activity. Its presence is more likely to occur during the late spring / summer than during the winter and allows easy access to the land beyond the horizon on VHF.

The F layer

The F layer is the outermost of our ionosphere and it begins at an altitude of about 130km divided into F1 and F2. It is not possible to clearly identify a top limit as it is very variable and often with a gradient border. Thanks to its high reflectivity, the F layer is the most important for long distance communications from a few MHz up to tens of MHz.

The layer F1 is the area located between 130 and 210km altitude where the electronic concentration is in the order of 200 billion electrons per cubic meter.

The F2 layer, the outermost of the ionosphere's layer, is where free electron concentration is generally the highest and very much dependent on the solar activity: its values are between 1000 billion electrons per cubic metre during summer daytime and "just" 50 billion electrons per cubic meter during the winter night. Modern scientific observations show that the electron concentration decreases in an approximately exponential way with the height. At 1000km the electron concentration is normally in the order of 10 billion electrons per cubic meter and is the limit where we usually consider the ionosphere finishes.

Propagation phenomena
In the amateur radio world several propagation phenomena are experienced due to the various layers of the atmosphere and the ionosphere with different impacts subject to the frequency of the radio waves. Without going into too much physics, let's examine the most common ones, which may also affect our satellite traffic.

Super refraction
- enhanced propagation of signals over the horizon because of atmospheric refraction. Particular conditions of humidity and atmospheric pressure gradient creating a layer called "conduits" which "guides" the signal beyond the horizon with minimum attenuation.
- bands involved: 145, 435, 1200, 2400, 5700, 10000MHz
- occurrence: very variable, mostly in summer, in areas with high pressure over the sea
- effects on earth-space communications: unknown, it is reasonable to believe in an increased range over the horizon with a slight extension of the time of acquisition of satellite

Atmospheric absorption
- the absorption of part of the radiation by some molecules (water and hydrogen, in particular) present in our atmosphere.
- bands involved: 1200, 2400, 5600, 10000, 24000MHz
- occurrence: continuous function of the relative humidity
- effects on earth-space communications: additional attenuation variable in function of the elevation of the satellite (because of the different thickness of the atmosphere to cross) and the conditions of the atmosphere (in particular, moisture)

Rain scattering (RS)
- the attenuation (loss) for scattering from hydrometeors (raindrops, hail, snow, etc). There are several mathematical models to estimate the attenuation due to rain, among others, which is worth mentioning:

Model	Minimum Freq [GHz]	Maximum Freq [GHz]
CCIR	3	100
CRANE	1	100
Medhurst	7.5	30
Ryde	4	13

- Given the deeply different mathematical approaches used by various studies and models, the phenomenon becomes more apparent as the frequency increases.
- bands involved: 1200, 2400, 5700, 10000MHz
- occurrence: variable, associated with rainfall, mainly storm
- effect on earth-space communications: severe increase of attenuation, spectral spreading and degradation of the polarisation coherence.

Tropospheric scattering
- propagation beyond the horizon of signals due to atmospheric refraction
- bands involved: 145, 435, 1200, 2400, 5700, 10000MHz

- occurrence: always present, although at very varying degrees
- effect on earth-space communications: slight increase of the communications range over the horizon with slight extension of the time of acquisition of the satellite.

Aurora Borealis (AU)

- The formation of ionised gas clouds in the upper atmosphere (100-400km) due to the interaction of the solar wind with the earth's magnetosphere, with fluorescent effects, visible as if the sky had turned into a huge neon lamp. The colours depend on the type of ionised gas in accordance with the table:

Colour	Ionised gas	Note
green	O_2	low altitude
yellow	O_2	rather faint
pearl white	O_2	faint
red	$O_2 + N_2$	shiny

- bands involved: 21, 24, 29, 145, 435MHz
- occurrence: many hours every year on sub-polar latitudes (50 degrees - 65 degrees N). Predictable in advance
- effect on earth-space communications: significant increase in noise, lock-out (total blockade of radio communications), corruption of spectral coherence of the signal, effects of scintillation

Meteor Scatter (MS)

The formation of ionised "stripes" due to the ablation of meteors against the atmosphere. The phenomenon is also known as "shooting stars" and this is its radio image. Every year the earth crosses many "meteor clouds". The most important ones from the point of view of radio communications are mentioned below, ranked according to the hourly rate provided:

Name	Date	Average peak date
Quadrantids	1st to 5th January	3rd January
Geminids	7th to 17th December	14th December
Perseids	17th July to 24th August	12th August
Age Aquarids	19th April to 28th May	6th May
Arietidi	22nd May to 2nd July	7th June
South Delta Aquarids	12th July to 19th August	28th July
Orionids	2nd October to 7th November	21st October
Lyrids	15th to 28th April	22nd April
Ursidi	17th to 26th December	22nd December

- bands involved: 145-435MHz
- occurrence: low, some days a year. Moment of occurrence and intensity predictable with high accuracy
- effect on earth-space communications: occasional brief disturbance from other services and extension of the operating range

The Ground-Space-Ground Propagation

Sporadic E (ES)
Local and sudden formations of high density electronic clouds (about 500 billion of free electrons per cubic meter, up to 12 times higher than average) in the E layer of the ionosphere, around 110-120km altitude.
- bands involved: 21, 24, 29, 145MHz
- occurrence: more frequent from May to August and from 7am to 9pm local time, for an average station at 45 degrees latitude. We may consider a few hours a day on 21 and 29MHz and few hours a year on 145MHz.
- effect on earth-space communications: increased noise from other terrestrial services, lock-out (total block of the radio communication), extended range beyond the horizon up to about 2400km.

Transequatorial Anomaly (TEP)
An anomaly on the ionosphere that allows connections far beyond the horizon between stations located in the same meridian and latitude "symmetrically placed" around the equator.
- bands involved: 21, 24, 29, 50, 70 and 145MHz
- occurrence: very rare
- effect on earth-space communications: increased noise from other terrestrial services, lock-out and extending the range beyond the horizon for many thousands of miles

Field aligned irregularity (FAI)
- different phenomena are grouped under this name, many of which are not yet fully understood. It appears, however, that the FAI anomaly occurs in common situations around the end of an Es or Aurora phenomenon and it appears to be due to the ionisation of the E layer
- bands involved: 21, 24, 29, 50, 70, 145, 435MHz
- occurrence: low, usually a few hours a year, often correlated with ES or Aurora
- effect on earth-space communications: severe signal distortion, abnormal Doppler Effect.

Ionospheric reflection (F2)
Ionisation of the F layer at high altitudes due to solar activity
- bands involved: 21, 24, 29MHz
- occurrence: very variable, from almost nothing up to almost constant (depending on the solar activity /sunspots number)
- effect on earth-space communications: increased noise from other terrestrial services, lock-out, extending of the range beyond the horizon for many thousands of miles

Faraday rotation
Due to the interaction of the radio signal and the ionised layer, the wave plane is rotated thereby changing its polarisation. The phenomenon worsens at low frequencies and is practically negligible at frequencies higher than a GHz or so while it increases at low elevation angles. The following table provides an indicative number of complete rotations of the wave plane that a generic signal coming from space may suffer.

	29MHz	145MHz	435MHz
Zenith (90 degrees)	24	1.4	0.25
Horizontal (0 degrees)	80	44	0.4

Hamsat - Amateur Radio Satellites Explained

- bands involved: 21, 24, 29, 145, 435MHz
- occurrence: always present which worsens with increased solar activity
- effect on earth-space communications: signal fading due to mismatching between wave plane and receiving antenna polarisation (it applies only in the case of linear polarisation).

And here is a summary table of propagation modes which are associated with each band used in our traffic-earth space:

Band											
	24GHz		■		■						
	10GHz		■		■						
	5.7GHz		■		■						
	2.4GHz			■	■						
	1.2GHz				■						
	435MHz				■	■	■				
	145MHz					■	■	■			
	29MHz							■	■		■
	21MHz							■	■	■	
	Phenomena altitude	< 1km	< 5km	< 10km	< 10km	80-120km	80-120km	90-20km	90-1000km	95-130km	130-400km
		Super refraction	Atmospheric absorption	Rain scatter (RS)	Tropospheric scattering	Aurora Borealis (AU)	Meteor scatter (MS)	Sporadic E (ES)	Transequatorial Anomaly (TEP)	Field aligned irregularity (FAI)	Ionospheric reflection (F2)
	Phenomena related to radio wave propagation										

The Link Budget

Link-budget, is the ability of a system to ensure a radio link, or to evaluate its performance under varying conditions. In amateur satellite communications, signals are transmitted to the satellite as well as received back on the earth. Since the power on board the satellite is limited, the space to ground segment is certainly the most critical, where we have to receive signals transmitted with minimal power and which experience interference from other services on the ground.

The quality of the result depends on many factors. Below we analyse them one by one (with some indication of typical figures):

- P_{tx} transmitter power on board the satellite: from fractions of a W to a few W
- L_{tx} transmission loss from transmitter to its antenna: fractions of dB
- G_{tx} satellite antenna gain: a few dB
- F   frequency of the transmission: it may range from 21MHz up to 24GHz
- D   distance of satellite to receiving station: from 300 to 60000km
- G_{rx} receiving antenna gain: 0 to 20dB typical
- T_{coax} temperature of the transmission line from the antenna to the receiver: 290°C typical
- L_{coax} line antenna to receiver losses: from almost 0dB to some dB
- NF   noise figure of the receiver: from a few tenths to few dB
- T_{ant} equivalent temperature of the receiving antenna: from tens to thousands of K
- BW   bandwidth of the communication channel: from tens of Hz to a few MHz
- PL   signal attenuation of the space to ground path

Now let's see the equations that govern the links between the variables listed above and the result achievable (signal/noise ratio).

Given the relative complexity of the general formula, it is better to break the analysis into three main parts related to the transmitting station, radio link and receiving station.

Transmitter

Let's introduce a very convenient quantity: the ERP.

ERP stands for *Effective Radiated Power* or *Equivalent Radiated Power*.

It is a standardised theoretical measurement of the level of radio frequency energy in Watts (using SI units) which is determined by adding algebraically all the losses and gains of the transmitting system.

The ERP includes therefore the power of the transmitter (Transmitter Power Output, TPO), the losses of the transmission line between the transmitter and the antenna, and the antenna gain.

The ERP is defined as the product of transmitter power (net of losses of transmission lines) and the antenna gain.

If all the variables are expressed in dB, the formula becomes:

$$ERP = P_{tx} + L_{tx} + G_{tx}$$

Example:
- P_{tx} = 0dBW (1W)
- L_{tx} = -0.5dB
- G_{tx} = 2dB
- **ERP** = 0 + (-0.5) + 2 = 1.5dBW (1.41W)

In this example, our transmitter will produce at a given distance and in the direction of the main lobe of the antenna, the same electromagnetic field that would have been produced using a transmitter of 1.41W connected via a low-loss transmission line to a lossless isotropic radiator placed in the same position of our transmitting aerial.

The Path

In this discussion, the space between the transmitter and receiver is considered non-dissipative in respect of the transmitted energy. But it is common place to attribute an attenuation which represents the amount of signal transmitted but not collected by the receiving antenna.

Let's see the equation of what is commonly called the path loss.

$$PL = 32.45 + 20 * \log(D) + 20 * \log(F)$$

With D (the path length) in km and F (frequency of the link) in MHz, *PL* is measured in dB.

It should be noted that this value is valid in free space which does not take into account the dispersive and dissipative phenomena that may occur in the atmosphere and ionosphere. More details on this additional signal loss are discussed in the chapter on the earth-space propagation.

The Receiving Station

On the ground, in our amateur radio station we have to deal with antennas, cables, receivers and noise (unfortunately). All these factors determine the capability of our station and, in more technical words, determine the total noise power (P_n), a parameter that identifies the minimum actually receivable signal. The physical sense of this quantity is to indicate the equivalent temperature of a resistor connected to the place of the antenna in front of an ideal receiver (without noise) which would produce the same "masking" of the incoming signals due to the noise generated by it.

Here again, let's divide the problem by assessing the 'equivalent temperature of the system' (TSYS, expressed in K), a parameter that measures the sensitivity of a receiver regardless of the bandwidth and the antenna gain:

$$T_{sys} = T_{ant} + T_{coax} * (L_{coax} - 1) + L_{coax} * T_{rx}$$

Where:

- L_{coax} attenuation of the cable between antenna and the receiver in a pure number (not in dB).
- T_{rx} equivalent temperature of the receiver. More frequently amateurs use the noise figure quantity which is linked to Trx as:

$$T_{rx} = 10^{\frac{NF}{10}} * T \text{ with } T = 290°K \text{ as default}$$

We can now combine all the elements and calculate the *Pn* mentioned above:

$$P_n = 10 * \log(B) + 10 * \log(T_{sys}) - 228.6 - G_{rx}$$

The Link Budget

Now, after knowing all the factors, we can calculate the expected signal/noise ratio, SNR achievable by our system:

$$SNR = ERP - P_n - PL$$

Compared with the real case, the calculated value is often optimistic and is to be regarded as a "best-case" because we have not considered the following effects:
- specific local disturbances problems (to be considered in Tant)
- the space-to-ground path is sometimes dissipative or dispersive (especially at high frequencies)
- the satellite antenna is not pointing straight at the receiving station (squint angle)
- the polarisation of the receiving antenna does not fully match the one of the incoming signal (almost always!)

Now let's see some practical examples related to the most typical ham situations:

29MHz down link, dipole, SSB, LEO satellite

System Data

P_{tx}	1	W	transmitter power
G_{tx}	0	dBi	transmitting antenna gain
F	29	MHz	frequency
D	1300	km	path length
G_{rx}	0	dBi	receiving antenna gain
T_{coax}	300	K	receiving transmission line temperature
L_{coax}	0.5	dB	transmission line losses
NF	6	dB	receiver noise figure
T_{ant}	5000	K	equivalent antenna temperature
BW	3000	Hz	receiver bandwidth

Results

P_{tx}	0.0	dBW	transmitter power,
ERP	0.0	dBW	ERP transmitter power,
PL	124	dB	path loss
L_{coax}	1.12		transmission line losses (ratio)
T_{rx}	865	K	receiver noise equivalent temperature
T_{sys}	6007	K	equivalent noise temperature of the system
P_n	-156.0	dBW	equivalent total noise power at the receiver
SNR	32.1	dB	signal to noise ratio achievable

145MHz downlink, 6el–yagi, FM, LEO satellite at apogee

System Data

P_{tx}	0.2	W	transmitter power
G_{tx}	1	dBi	transmitting antenna gain
F	145	MHz	frequency
D	800	km	path length
G_{rx}	11	dBi	receiving antenna gain
T_{coax}	300	K	receiving transmission line temperature
L_{coax}	1	dB	transmission line losses
NF	3	dB	receiver noise figure
T_{ant}	1000	K	equivalent antenna temperature
BW	15000	Hz	receiver bandwidth

Results

P_{tx}	-7.0	dBW	transmitter power
ERP	-6.0	dBW	ERP transmitter power
PL	134	dB	path loss
L_{coax}	1.26		transmission line losses (ratio)
T_{rx}	289	K	receiver noise equivalent temperature
T_{sys}	1441	K	equivalent noise temperature of the system
P_n	-166.3	dBW	equivalent total noise power at the receiver
SNR	26.5	dB	signal to noise ratio achievable

436MHz downlink, 15el yagi, FM data, LEO satellite at apogee

System Data

P_{tx}	0.5	W	transmitter power
G_{tx}	2	dBi	transmitting antenna gain
F	436	MHz	frequency
D	800	km	path length
G_{rx}	13	dBi	receiving antenna gain
T_{coax}	300	K	receiving transmission line temperature
L_{coax}	2	dB	transmission line losses
NF	2	dB	receiver noise figure
T_{ant}	150	K	equivalent antenna temperature
BW	20000	Hz	receiver bandwidth

Results

P_{tx}	-3.0	dBW	transmitter power
ERP	-1.0	dBW	ERP transmitter power
PL	143	dB	path loss
L_{coax}	1.58		transmission line losses (ratio)
T_{rx}	170	K	receiver noise equivalent temperature
T_{sys}	594	K	equivalent noise temperature of the system
P_n	-170.8	dBW	equivalent total noise power at the receiver
SNR	26,5	dB	signal to noise ratio achievable

Hamsat - Amateur Radio Satellites Explained

2400MHz downlink, 60cm dish, SSB, HEO satellite

System Data					**Results**			
P_{tx}	1	W	transmitter power		P_{tx}	0.0	dBW	transmitter power
G_{tx}	5	dBi	transmitting antenna gain		ERP	5.0	dBW	ERP transmitter power
F	2400	MHz	frequency		PL	188	dB	path loss
D	25000	km	path length		L_{coax}	1.12		transmission line losses (ratio)
G_{rx}	20	dBi	receiving antenna gain		T_{rx}	120	K	receiver noise equivalent temperature
T_{coax}	300	K	receiving transmission line temperature		T_{sys}	251	K	equivalent noise temperature of the system
L_{coax}	0.5	dB	transmission line losses		P_n	-190.3	dBW	equivalent total noise power at the receiver
NF	1.5	dB	receiver noise figure		SNR	7.,3	dB	signal to noise ratio achievable
T_{ant}	80	K	equivalent antenna temperature					
BW	2700	Hz	receiver bandwidth					

436MHz downlink, 15el yagi, SSB, HEO satellite at apogee

System Data					**Results**			
P_{tx}	1	W	transmitter power		P_{tx}	0.0	dBW	transmitter power
G_{tx}	2	dBi	transmitting antenna gain		ERP	2.0	dBW	ERP transmitter power
F	436	MHz	frequency		PL	173	dB	path loss
D	25000	km	path length		L_{coax}	1.26		transmission line losses (ratio)
G_{rx}	13	dBi	receiving antenna gain		T_{rx}	170	K	receiver noise equivalent temperature
T_{coax}	300	K	receiving transmission line temperature		T_{sys}	441	K	equivalent noise temperature of the system
L_{coax}	1	dB	transmission line losses		P_n	-180.8	dBW	equivalent total noise power at the receiver
NF	2	dB	receiver noise figure		SNR	9.6	dB	signal to noise ratio achievable
T_{ant}	150	K	equivalent antenna temperature					
BW	2700	Hz	receiver bandwidth					

2400MHz downlink, 1m dish, low speed DVBS, ISS overhead

System Data					**Results**			
P_{tx}	5	W	transmitter power		P_{tx}	7.0	dBW	transmitter power
G_{tx}	6	dBi	transmitting antenna gain		ERP	13.0	dBW	ERP transmitter power
F	2401	MHz	frequency		PL	156	dB	path loss
D	600	km	path length		L_{coax}	1.06		transmission line losses (ratio)
G_{rx}	16	dBi	receiving antenna gain		T_{rx}	59	K	receiver noise equivalent temperature
T_{coax}	300	°K	receiving transmission line temperature		T_{sys}	150	K	equivalent noise temperature of the system
L_{coax}	0.25	dB	transmission line losses		P_n	-148.9	dBW	equivalent total noise power at the receiver
NF	0.8	dB	receiver noise figure		SNR	6.2	dB	signal to noise ratio achievable
T_{ant}	70	K	equivalent antenna temperature					
BW	2.5	MHz	receiver bandwidth					

The Ground Station

When using the term "ground station" we mean all the equipment and systems placed at a site on the earth which is intended to carry out satellite communications. In fact we are talking about our amateur radio station at home.

The components
The necessary components can be highly variable depending on the communication to be carried out and available economic resources. However, two elements are essential: the antenna and a receiver.

The antenna rotator
Its task is to guide the antennas in the direction of the satellite in its continuous apparent motion in the sky.

Let's look at the main features related to the choice.

Braking torque
Each antenna system, even when at rest, is subject to forces that tend to make it rotate. Among them we can name the imbalanced masses and especially the action of the wind. The braking torque, thus, indicates the maximum resistance (torque) that the rotator can provide against the rotation. This value depends very much on the braking system chosen by the manufacturer and by its sizing.

Basically there are four types:

Type	Mechanical description	Example	Pro	Cons	Drawing
No brake	It uses the natural braking torque that presents a chain of reduction gears	Yaesu G-250	Affordable, light and compact	Low braking capability	
Disk brake	Similar idea to the automotive version: a disc on the input shaft pressed against a friction surface when the rotator is stationary	Kempro KR-600	Affordable and space saving	Suitable only for small values of braking which will be worn out over time	
Wedge brake	A wedge is inserted into a fluted crown when the rotator is stationary	HAM-IV	Very effective	Noisy. It works only in certain positions (ie every 6 degrees)	
Worm gear	Based upon the irreversibility of the worm gear	Create RC5B	Very good values of braking	Cumbersome and expensive	

In the case of us exceeding the braking torque, the antenna system will not remain in the set position thus preventing the tracking of the satellite and exposing itself to the potential risks for the safety of the installation and perhaps of the people. It is also worth noting that the whole structure of the rotator is sized proportionally to the braking torque indicated.

PC Interface

Many rotators are designed to be interfaced with a personal computer. This feature allows for example the automatic tracking of the satellites. Some models have an output already compatible with a PC others have only a connection for PC.

Accuracy / play

From these parameters, we can identify the beaming indeterminacy, useful to assess the suitability of a rotator to steer narrow lobe antennas, typical of higher frequencies. Precision is usually defined by the type of installed position sensor while the play depends on the configuration and the quality of the mechanical gears used.

As a rule of thumb, we can accept a precision/play of up to half of the lobe (-3dB) of the most critical antenna. As a general guideline, the amateur grade rotators have values in the range of 1~3 degrees when brand new and well balanced.

Angle of rotation

All the azimuthal rotators have the ability to rotate for at least 360 degrees, allowing themselves to explore all the directions of the horizon. In recent years, some models have become popular with rotation up to 450 degrees. With those rotators it is possible to continue the tracking without having to wait for the repositioning to the opposite end of the stroke.

The elevation rotators generally have a run of 90 degrees or 180 degrees. The latter allows tracking manoeuvres faster in case of particular trajectories (passing almost overhead). But not all the tracking software offers this feature.

Speed

Normally it indicates the time taken to rotate the output shaft of 180 degrees or 360 degrees. The typical value is in the range of one minute. Fast rotors allow you to move quickly between different points but stress the mechanical system due to the high acceleration. Complex antenna systems are generally combined with "slow" rotors which offer limited mechanical stress and better accuracy. Variable speed drive systems are rare in the amateur world mainly as a result of self-construction.

Torque

It indicates the ability to rotate the antennas. The torque required is generally proportional to the unbalance of the masses (in the case of the elevation in particular), the aerodynamic load (in the presence of wind) and the moment of inertia of the system.

Vertical load

It indicates the maximum weight that may be vertically placed on the structure of the rotator without compromising its reliability and functionality. You can increase it by combining the rotator with one or more external bearings which are placed in a special structure and commonly called "cage".

Number of wires

It indicates the number of cables required to drive the rotator and the brake, and reports back to the controller in the shack the position of the antenna. They range from the simplest 3-wire (TV rotor type) to the most complex 6 and 8 wires.

Let's see now, just as example, some typical values of a number of rotators popular in the amateur market.

Brand	Model	Rotation [degrees]	Speed [sec*360 degrees]	Torque [kgm]	Braking [kgm]	Mechanical configuration	Motor	Weight[kg]	Vertical load [kg]	N° of wires
Alinco	EMR-400	365	60	5.5	15	-	-	5	200	-
Create	RC5-1	360	-	5.9	70	Gear chain + worm gear	AC induction, reversible	-	400	-
Create	RC5A-2	360	-	16	150	Gear chain + worm gear	AC induction, reversible	-	700	-
Create	RC5B-3	360	-	22.1	200	Gear chain + worm gear	AC induction, reversible	-	700	-
Daiwa	DR-7500R	-	-	-	20	4 motors, modular	AC induction, reversible	5.5	200	6
Hy-gain	CD 45-II	-	-	6.9	9.2	Gear chain	AC induction, reversible	10	-	7
Hy-gain	HAM-IV	-	-	9.2	58	Gear chain + wedge	AC induction, reversible	10.8	-	8
Hy-gain	T2X	-	-	11.5	103	Gear chain + wedge	AC induction, reversible	12.7	-	8
Hy-gain	HDR-300A	-	-	-	-	Gear chain + wedge	AC induction, reversible	24.9	-	8
Kenpro	KR 2000RC	-	-	-	100	-	AC induction, reversible	9	250	8
Kenpro	KR 500	180	148	-	20	Gear chain	AC induction, reversible	3.3	400	6
Kenpro	KR 600 RC	-	-	6	40	Gear chain + disk brake	AC induction, reversible	4.6	200	6
Yaesu	G-500A	180	122	10	20	Gear chain	AC induction, reversible	-	200	6
Yaesu	G-450XL	450	51	5	30	Gear chain	AC induction, reversible	-	110	6
Yaesu	G-800S/SDX	450	55 / 43-93	5.9 - 10.9	40	Gear chain	DC motor	-	200	5
Yaesu	G-1000SDX	450	43 - 93	5.9 - 10.9	59.8	Gear chain	DC motor	-	200	5
Yaesu	G-2800SDX	450	60 / 190	8 - 25	250	Gear chain	DC motor	-	300	6
Yaesu	G-5600B hor.	-	53	6	20	Gear chain	AC induction, reversible	-	200	2x6
Yaesu	G-5600B vert.	-	116	14	40	Gear chain	AC induction, reversible	-	200	2x6
Yaesu	G-250	360	43	2	6	Gear chain	AC induction, reversible	-	50	6

Preamps

It is the first stage of the reception chain to which most of the sensitivity of the system is assigned. As well, it is one of the components of the amateur station where self-construction and innovative development solutions are frequent.

Let's Look at the Main Features

Noise Figure

The noise figure, often abbreviated as NF, is a quantity used in telecommunications together with the equivalent noise temperature to quantify the internal noise of a device.

In a general sense, the noise figure quantifies the deterioration of the signal-to-noise ratio (SNR) due to their intrinsic noise as:

$$F = \frac{SNR_{inp}}{SNR_{out}}$$

where F is the noise figure expressed as a pure number. The NF commonly used is calculated from F as:

$$NF = 10 \log(F)$$

The noise figure is linked to the equivalent noise temperature by the relation:

$$F = \frac{T + T}{T} \quad \text{and vice versa:} \quad T_r = (F - 1)T$$

where T is the absolute temperature (expressed in Kelvin) based on which F is calculated.

The values of F (or NF) provided by the manufacturers of electronic equipment is therefore to be considered as the reference temperature used for their calculation, conventionally of 290K it is not specified otherwise.

For a quick conversion between NF and T, we can employ the graph shown here (**Fig 4.1**), where along the radius you can read the temperature in K and along the circumference the NF values (expressed in dB).

For the space-to-ground traffic, this important parameter should always remain below 1dB (75K) for frequencies from 430MHz and above. Values of 0.5-0.7dB are the most cost-effective available today.

Gain

It indicates the relationship between the output and input signal level.

Low values, generally less than 12-14dB, are not recommended because they do not "compensate" enough for the losses in the cables and the receivers' noise.

In fact, in a chain of devices such as our receiving systems, the contribution of each stage to the total noise (F) is inversely proportional to the gain of the stage that

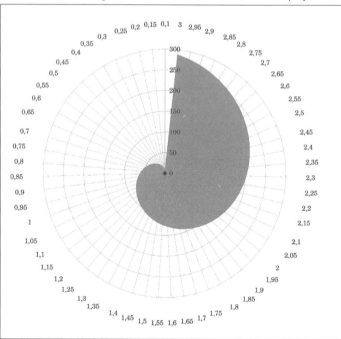

Fig 4.1: Conversion from NF to T noise

precede it as shown in the following formula, known as Friis formula from the name of its first extender:

$$F = 1 + \frac{\sum_{i=1}^{N} F_i}{\prod_{j=1}^{i} G_j}$$

We can easily rewrite as:

$$F = F_1 + \frac{F_2 - 1}{G_1} + \frac{F_3 - 1}{G_1 G_2} + \ldots + \frac{F_n - 1}{G_1 G_2 \ldots G_{(n-1)}}$$

Where Fn is the noise figure of stage n, and Gn is the gain (linear, not dB).
For those more accustomed to the equivalent noise temperature, the same formula becomes:

$$F = F_1 + \frac{F_2 - 1}{G_1} + \frac{F_3 - 1}{G_1 G_2} + \ldots + \frac{F_n - 1}{G_1 G_2 \ldots G_{(n-1)}}$$

Higher values of gain of the preamp, indicatively above 20 ~ 22dB, instead might limit the dynamics of the system and the general performance of many receivers. Typical values suitable for most of the situations are therefore in the range of 16-22dB.

Below is an example:

Preamplifier noise figure	1dB
Preamplifier gain	16dB
Coax cable losses	2dB
Receiver noise figure	4dB

This chain of elements has a global noise figure of 1.25dB or 97K equivalent noise temperature. Contribution of each element is shown in the graph: **Fig 4.2**

Front-end bandwidth

There are situations in which a normal front-end may fall into trouble, such as when managing J (V / U) and L/S modes. The common problem in both of the situations is, that at the input of the reception chain, a fraction of the transmitted signal is present (usually the 3rd harmonic) which can result in saturation of the active device of the front-end (or any protective/switching diodes at the input), thus deteriorating dramatically reception quality.

For example, in a typical amateur station operating in J mode, the antennas for 144MHz and 430MHz, parallel to

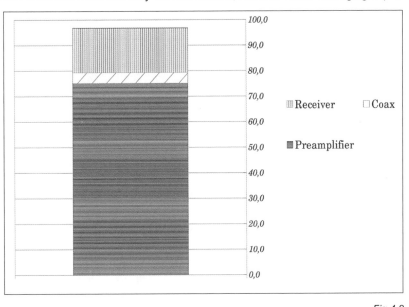

Fig 4.2

each other and spaced at approximately 3m, offer approximately 40dB of isolation. Assuming we transmit 50W on 2m we will find at the input of the 70cm preamp about +7dBm or if you like more, some 5mW!

Often the problem of poor reception is wrongly allocated to the harmonic emission of the transmitter. In fact: 145MHz x 3 = 435MHz or 1269MHz x 2 = 2538MHz

If this were the case, with a simple low-pass filter placed at the output of the transmitter the problem would be solved. This test could be used as a discriminator to address the problem correctly.

By way of explanation, let's see some calculations inspired by real-life (2m/70cm).

Situation 1: RTX commercial, medium power, the antennas are spaced a few meters.

Situation 2: as 1, but introducing a low pass filter on the transmitter at 145MHz

Situation 3: RTX commercial with the addition of a linear amplifier of 1kW, filtered and not particularly well-spaced, antennas on independent support.

Situation 4: as 3, but adding a good filter on the output of the transmitter.

Test	Radio	Power @ 145MHz	3rd harmonic	Tx power on 3rd harmonic	145-435 antenna coupling	435-435MHz antenna coupling	3rd harmonic rx level	435MHz Rx, input level @ 145MHz
1	Commercial RTX	50W 47dBm	-50dB	0.5mW -3dBm	-40dB	-30dD	S9+40dB 'Smotor' -33dBm	500mV/50Ohm 7dBm
2	Commercial RTX + LPF	50W 47dBm	-85dB	1.6E-4mW -38dBm	-40dB	-30dB	S9+5dB 'Smeter' -68dBm	500mV/50Ohm 7dBm
3	Commercial RTX + 1kW PA	1000W 60dBm	-30dB	1000mW 30dBm	-65dB	-45dB	S9+58dB 'Smeter' -15dBm	126mV/50Ohm -5dBm
4	Commercial RTX + LPF	1000W 60dBm	-70dB	0.1mW -10dBm	-65dB	-45dB	S9+18dB 'Smeter' -55dBm	126mV/50Ohm -5dBm

Analysis:

Situation 1: the third harmonic of the transmitter, around 435-438MHz, although reaching the front end would not be the main cause of the trouble. In fact, at the preamp's input we will find a level of about 500mV of signal at 145MHz which makes the protection diodes clip the input (therefore creating a huge comb of spurious signals) and most likely saturate the first active devices. The reception is severely compromised.

Situation 2: the effect of the third harmonic of the transmitter is reduced slightly. The front end, however, continues to be saturated by the 500mV signal at 145MHz. So the reception does not improve appreciably compared to case 1.

Situation 3: the third harmonic of the transmitter is very high and probably will produce inter-modulation products across the UHF band. The 145MHz signal is not particularly intense and its presence is likely to be masked by the non-linear effects produced by the third harmonic.

Situation 4: the filtering intervention is clearly visible with immediately appreciable benefits. The third harmonic is reduced to a manageable level by the dynamics of most preamplifiers.

It is not only our own uplink signal which creates difficulties for the front end of the receiver. Also the surrounding environment often puts a strain on our receiver. For those who operate in the 430MHz, the main interference comes from TV transmitters in band IV and the civil service just outside the amateur band.

The Ground Station

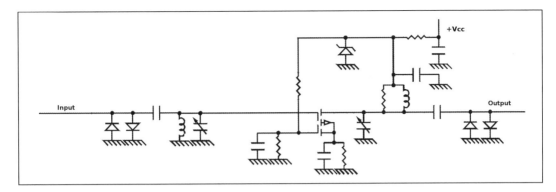

Fig 4.3: Example of a typical input stage of a commercial receiver or preamplifier

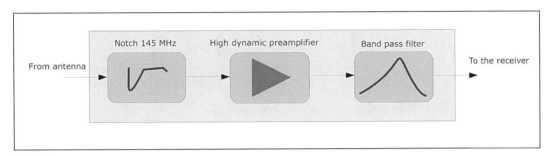

Fig 4.4: Block diagram of a well tailored front end for satellite operations

While for the S-band (240MHz) troubles may come from repeaters, audio and video senders, Wi-Fi, etc.

The problem so far analysed is common for different transceivers and most of the commercial preamplifiers. I do not want in any way to blame designers and retailers. It is worth noting that to design and build devices suitable for this specific use would cost more moneythan this small, tailored market would find acceptable. So it is logical for commercial products to meet the needs of the mainstream market without adding cost and performance that only a minority would be willing to pay for and use.

Having said that, let's see an example highlighting the typical weaknesses of a generic commercial rig based on a regular front-end: (**Fig 4.3**)

The diodes at the input (and sometimes at the output too) of the front end: they have a protective function for the active device. But in case of broad signals they enter into conduction, distorting the input signal and generating disturbances over the entire band.

The input is generally broad-band: this is for the sake of cost, construction simplicity and adaptation for best noise performance of the amplifier. Unfortunately, the input designed that way does not protect the active device from strong, out of band signals.

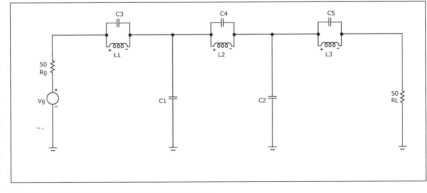

Fig 4.5: Notch input filter schematic diagram

A good way to solve the problem (example for the J or V/U mode) could be drawn in **Fig 4.8**.

The first block on the left is a notch filter (band-stop) tuned on the 2m band. It must possess good attenuation but also minimal loss at 435MHz so as not to penalise the total noise figure of the whole receiving chain.

The stage in the middle is a good preamplifier with high dynamics and low noise.

The third section is a narrow band-pass filter to eliminate the signals which are adjacent to the band in use.

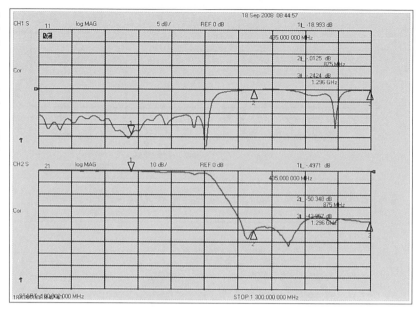

Fig 4.6: Measured response of the notch filter

'A practical realisation of the notch may be the one proposed' by Heinz Hildebrand DL1CF in an old article*.

The circuit diagram, as being imported into the simulation program MicroCap, is as follows (**Fig 4.5**):

The circuit is composed of three parallel resonant cells in series to the signal path. These said cells have high (theoretically infinite) impedance at their resonant frequency, causing a significant attenuation of the input signal. Above the resonant frequency, the three cells behave as a series of inductances. The two parallel capacitors connected to the ground are then used to compensate for the inductive reactance at the frequency of listening while they are virtually negligible at 144MHz.

The original article stated the possibility of obtaining the attenuation up to 80dB at 145MHz and only 0.1dB at 435MHz. (**Fig 4.6**)

The measures on some prototypes confirm this expectation showing the attenuation at 145MHz of at least of 60dB, therefore ensuring excellent immunity against the transmitter's signal

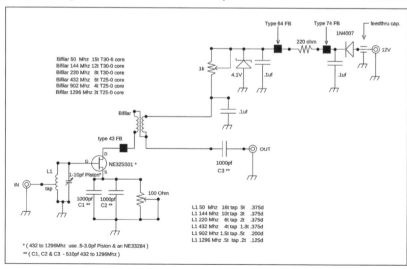

Fig 4.7: High dynamic and low noise preamplifier designed by WA2ODO based upon NE325S01 device

- attenuation on the 70cm band is only 0.3dB, input and relay connections included
- the return-loss in the UHF band is very good, at least 20dB for the 150MHz band: this facilitates the matching for best noise performances of the preamplifier and its stability

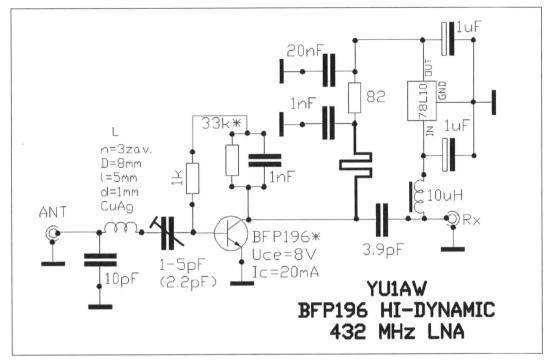

Fig 4.8: High dynamic and low noise preamplifier designed by YU1AW based upon BFP196P

- Similarly, you can approach the front end of the S-band receiver which has to manage the return signal at 1268MHz uplink and maybe also some various local disturbance sources.

Anyhow, the notch will do very little if it is not followed by a good preamp with high dynamics and low noise.

In recent years, the availability of new components and systems for computer-aided design has made it much easier to solve the problem of designing and producing preamplifiers with very low noise and with good dynamics at the same time. Purely by way of example, it is worth mentioning various excellent work of YU1AW and WA2ODO based on devices such as MGF1801, MGF0904A, BFP196P or BFG540X.

Coaxial cables

Coaxial cables and connectors are crucial to the success of our project, but unfortunately are not considered as often as they should be. A poor choice or deterioration can lead to severe performance loss, often difficult to diagnose. Their task is to convey the signal from the transmitter to the antenna and from the antenna in the opposite direction towards the receiver, with minimum loss.

"Quality" of the cables must be considered when choosing:

Bending radius varies from a few mm for thinner cable to several dm for cables like cellflex of larger cross-section cables.

Maximum power: it indicates the maximum handling power in a safe condition for the cable. The value decreases rapidly with the working frequency due to increased loss. This data is usually expressed at 20°C of ambient temperature. If the cable is exposed to higher temperatures (for example outdoor along the mast/tower) it is necessary to further reduce the value indicated.

- Working temperature: generally from -20°C to 85°C, values suitable for most amateur applications. There are versions with PTFE insulation and sheathing specified up to 200°C intended for special high power applications (internal wiring of big power amplifiers) or passages in hot environments (ie chimneys, etc)
- Possibility of burial: this quality is useful when you have to cross a garden (for instance) and you want to hide the cables. Professional cables and larger cross-section coax usually can be buried directly into the soil.
- Speed factor: the ratio between the wave speed in the cables compared with the one in free space, usually varying between 66 and 86%. It is necessary to consider it only if the electrical length of the cable is known.
- Impedance: 50 Ohm typical. There are also versions at 25, 75, 95 and 125Ohms which are for special applications.
- Loss: each cable, no matter how well it is built, dissipates the heat part of the signal that it carries due to two phenomena: ohmic loss in the conductors and dielectric loss in the insulation. Both phenomena are proportional to the frequency and overall loss grows roughly with the square root of the frequency.

Among all the parameters seen, usually the most considered are the losses.

It should be noted that while a loss of transmitting power can be "recovered" by increasing the output power (often at a high cost!), during receiving what is lost from the cable will be permanent and it is not possible to recover it in any way.

That is why it is necessary to minimise the loss between the antenna and the first stage of the receiver.

Following the calculation shown in the chapter on the link-budget, let's simulate two alternative situations: one with zero cable loss (receiver directly connected to the antenna) and the other with 2dB of cable loss.

436MHz, receiver directly connected to the antenna

P_{tx}	0.1W	Transmitter power	P_{tx}	-10dBW	Transmitter power
G_{tx}	2dBi	Tx antenna gain	ERP	-8dBW	ERP value
F	436MHz	Downlink frequency	PL	173dB	Path loss
D	25000km	Distance from satellite to ground station	L_{coax}	1-	Loss of coax cable (ratio)
G_{rx}	13dBi	Rx antenna gain	T_{rx}	59K	Equivalent receivers noise temperature
T_{coax}	300K	Temperature of the coax cable	T_{sys}	179K	Equivalent systems noise temperature
L_{coax}	0dB	Losses of the coax cable from antenna to receiver	P_{rx}	-184,8dBW	Rx noise floor
NF	0.8dB	Noise figure receiver	SNR	3.6dB	S/N ratio
T_{ant}	120K	Equivalent antennas noise temperature			
BW	2700Hz	Channel bandwidth			

436MHz, receiver connected to the antenna with 2dB of losses due to the coax

P_{tx}	0.1W	Transmitter power	P_{tx}	-10dBW	Transmitter power
G_{tx}	2dBi	Tx antenna gain	ERP	-8	dBW ERP value
F	436MHz	Downlink frequency	PL	173dB	Path loss
D	25000km	Distance from satellite to ground station	L_{coax}	1,58	Loss of coax cable (ratio)
G_{rx}	13dBi	Rx antenna gain	T_{rx}	59K	Equivalent receivers noise temperature
T_{coax}	300K	Temperature of the coax cable	T_{sys}	388K	Equivalent systems noise temperature
L_{coax}	2dB	Losses of the coax cable from antenna to receiver	P_{rx}	-181.4dBW	Rx noise floor
NF	0.8dB	Noise figure receiver	SNR	0.2dB	S/N ratio
T_{ant}	120K	Equivalent antennas noise temperature			
BW	2700Hz	Channel bandwidth			

Difference: 3.6 - .02=3.4dB!

The Ground Station

The result may seem to be surprising that the change of only 2dB of loss will make the listening quality become 3.4dB worse!

Let's now see another case. With a higher frequency such as a down-link to 2400MHz where the converter/preamplifier is connected alternately directly on the feeder of the satellite dish, or about two metres away with a good cable.

2401MHz, receiver directly connected to the antenna

P_{tx}	0.5W	Transmitter power	P_{tx}	-3dBW	Transmitter power
G_{tx}	5dBi	Tx antenna gain	ERP	2dBW	ERP value
F	2401MHz	Downlink frequency	PL	188dB	Path loss
D	25000km	Distance from satellite to ground station	L_{coax}	1-	Loss of coax cable (ratio)
G_{rx}	20dBi	Rx antenna gain	T_{rx}	59K	Equivalent receivers noise temperature
T_{coax}	300K	Temperature of the coax cable	T_{sys}	99K	Equivalent systems noise temperature
L_{coax}	0dB	Losses of the coax cable from antenna to receiver	P_{rx}	-194.3dBW	Total receiving noise power
NF	0.8dB	Noise figure receiver	SNR	8.3dB	S/N ratio
T_{ant}	40K	Equivalent antennas noise temperature			
BW	2700Hz	Channel bandwidth			

2401MHz, receiver connected to the antenna with 0.6dB of losses due to the coax

P_{tx}	0.5W	Transmitter power	P_{tx}	-3dBW	Transmitter power
G_{tx}	5dBi	Tx antenna gain	ERP	2dBW	ERP value
F	2401MHz	Downlink frequency	PL	188dB	Path loss
D	25000km	Distance from satellite to ground station	L_{coax}	1.15-	Loss of coax cable (ratio)
G_{rx}	20dBi	Rx antenna gain	T_{rx}	59K	Equivalent receivers noise temperature
T_{coax}	300K	Temperature of the coax cable	T_{sys}	152K	Equivalent systems noise temperature
L_{coax}	0.6dB	Losses of the coax cable from antenna to receiver	P_{rx}	-192.5dBW	Rx noise floor
NF	0.8dB	Noise figure receiver	SNR	5.4dB	S/N ratio
T_{ant}	40K	Equivalent antennas noise temperature			
BW	2700Hz	Channel bandwidth			

Difference: 8.3-5.4=1.9dB!

Also in this case, the result clearly indicates that a small loss in the coax cable will generate far more impact on the final result.

This behaviour may appear singular due to the fact that the actual sensitivity of our reception chain is fixed by the temperature of the system as a sum of many contributions. The lower the temperature of the antenna and the receiver, the greater will be the "weight" on the sum of those due to the cable and any other devices in between (filters, circulators, coax connectors, etc).

To be better convinced, when in doubt it is sufficient to make a graphical representation as follows:

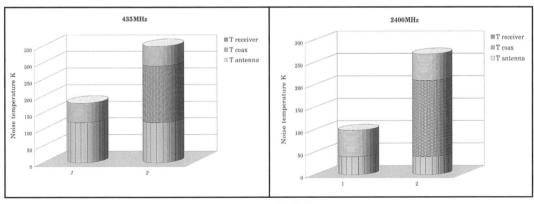

Fig 4.9   Fig 4.10

Graphical representation of the three main contributions to the total receiver's noise temperature as discussed in the example above

For amateur radio, there are basically five families of coaxial cables to be considered, depending on their diameter:

2-3mm: very flexible, suitable for small signal wiring in confined spaces.

5-6mm: this family, of which probably the RG58 is the most famous member, is mainly intended for indoor use, short distances and low power. The small diameter and high flexibility make it suitable for application in confined spaces. However, the attenuation introduced is significant and almost always unacceptable for long lines. The versions with PTFE (Teflon ®) dielectric could also withstand high power and find applications in power wiring such as a balun for instance.

10-11mm: these cables are the most common among amateur radio enthusiasts. The original of these cables was certainly the famous RG8/RG213. They are characterised by a good compromise between size, loss and flexibility, therefore well compatible with the home environment. In recent years, various versions with low dielectric losses have been developed, but unfortunately at the expense of flexibility. In general, they remain a good choice for connection that is not particularly long.

14-16mm: this is the family that "links" the amateur world with the professional one. The best known is probably the Cellflex cable, readily available as "scrap" from professional installations. Its performance is very good, for the price, but its considerable rigidity can make it complicated to use in confined spaces. There are versions that are more flexible, usually 13.5mm in diameter, which are probably the best compromise between electrical and mechanical performance.

18-30mm: these are the best low loss cables. Unfortunately cost and difficulty of installation due to the high rigidity limits their use in the amateur field. They are, however, essential when runs become very long and the frequency rises.

Let's now have a look at a summary table of cable loss for the most popular models. Data are expressed in dB/100m and are an average of the values declared by different manufacturers/retailers. In fact, there are indeed many cables on the market, often similar to each other even if under a different brand/name. It is advisable, in this case, to check the manufacturer's documentation for the updated data.

Model	Dia [mm]	21MHz	29MHz	145MHz	435MHz	1296MHz	2400MHz
RG178	1.9	23.5	26.2	53	95	176	244
RG179	2.54	20.5	22	36	55	89	121
RG188	2.79	22.4	23.8	38	62	114	166
RG187	2.79	20.5	22	36	55	89	121
RG174	2.8	16.3	18.2	36	59.8	110	169
RG303	4.32	5.5	6.4	15.4	27.7	51	74
RG142	4.95	5.5	6.4	15.4	27.7	51	73
LMR200	4.95		5.9	13.1	23.0	41.0	
RG58	5	6.3	7.5	18.8	35.2	69	100
RG58 FOAM	5	5.3	6	13.6	24.8	53	92
AIRCELL 5	5			11.8	20.9	38	53
RG302	5.23	4.9	5.5	12.6	24.2	49	72

Model	Dia [mm]	21MHz	29MHz	145MHz	435MHz	1296MHz	2400MHz
RG223	5.38	6.4	7.6	19.2	34.5	62	83
H155	5.4	4.2	4.9	11.2	19.8	37	54
LMR240	6.1		3.3	9.8	17.4	31.0	
RG8x FOAM	6.3	5	5.8	14	26.4	59	106
AIRCELL 7	7	3.1	3.5	7.9	14.1	26.1	39
Cellflex 1/4", SCF14-50	7.8	2.6	3.2	7.2	12.7	22.0	32.0
H100 / H200	9.8	1.8	2.1	4.9	8.8	16	21.4
Cellflex 1/4", LCF14-50	10	1.9	2.3	5.2	9.1	15.7	22.3
Cellflex 3/8" SCF38-50	10.2	1.9	2.3	5.2	9.2	15.9	22.8
Ecoflex 10	10.2			4.8	8.9	16.5	23.1
RG9 / RG214	10.3	3.2	3.7	9	17.1	34	51
RG8 / RG 213	10.3	2.6	3	7.5	14.8	30.4	45
RG225	10.3	2.3	2.6	6.1	11.6	22.7	33
RG213 FOAM	10.3	2.6	2.9	6	11	22	32.7
9913	10.3		2.0	5.6	10.2		
LMR400	10.3		2.6	5.3	9.8	17.4	
H1000 / H2000	10.3	1.7	2	4.8	8.5	15.7	-
CT50/20 INFLEX	10.3	1	1.3	4.8	9.6	17.3	25.2
AIRCOM PLUS	10.8	1.5	1.8	4.5	8.2	15.2	21.4
Cellflex 3/8", LCF38-50	11.2	1.5	1.9	4.2	7.4	12.8	18.2
LMR500	12.7		1.8	4.0	7.1	13.1	
Cellflex 1/2" SCF12-50	13.7	1.5	1.8	4.1	7.3	12.7	18.2
Ecoflex 15	14.6			3.4	6.1	11.4	16
LMR600	15		1.4	3.2	5.6	10.5	
Cellflex 1/2", LCF12-50	16.2	1.0	1.2	2.7	4.7	8.1	11.6
RG17 / RG218	22	1.2	1.5	4	8.2	17.3	25.1
Cellflex 7/8" UCF78-50	27.48	0.5	0.7	1.5	2.7	4.7	6.8
Cellflex 7/8", LCF78-50	27.8	0.5	0.6	1.42	2.54	4.45	6.47

The connectors

As already mentioned for coaxial cables, connectors are also a key element to the success of our project. A poor choice or deterioration can lead to severe performance loss, often difficult to diagnose. Their task is to connect different sections of the transmission line (coaxial cable), ensuring minimal loss and constant impedance. Often, the connectors are installed outdoors and should therefore be protected from moisture and contaminants.

In the table below, are some connectors most commonly used by radio amateurs with some relevant historical notes and use tips.

Picture	Connector	Notes	Max usable frequency	Notes
	BNC	Bayonet Navy Conn. or Bayonet Neil Concelman (inventor)	1 to 3GHz	Very low SWR connector developed in the 1940s for military purposes. Its limitation in frequency is due to the "unstable" contact of the inner pin and the bayonet coupling but the impedance is good to 10GHz.
	N	Navy or Paul Neill Bell Lab. (inventor)	12 - 18GHz	Designed in the 1940s for military systems up 4GHz. It was the first connector capable of operating at microwave frequencies, later improved for applications up 12 or 18GHz. It's the connector more widely distributed in the amateur world, suitable for both internal and external application and also for medium power handling.
	TNC	Threaded N	11 – 12GHz	Excellent connector combines the quality of N frequency with the small size of BNC. Its design is elegant and precise. Not really widespread among radio amateur
	UHF	-	200 – 300MHz	Developed by Mr C Quackenbush (Amphenol) in the 1930s for use in radio frequency, maybe the first standard connector for RF. Definitely the most common one because of historical and commercial reasons.
	7-16	-	5 – 7GHz	Designed in Europe for broadcast systems, it has been universally adopted for applications in cellular base stations. Common on the surplus market, it is an excellent choice for the amateur.
	SMA	-	18-26GHz	Developed by Bendix, it was initially called 3mm (in the 1960s), certainly the most successful connector for microwaves.

The Ground Station

Receivers

Together with the antenna, the receiver is one of the two essential elements for each ground station. There are many models and solutions available in the market as well as examples of high-level amateur home brewing.

In general, the performance it needs to offer shall be:

Selectivity

The selectivity has to match the modulation to be received. As rule of thumb we can take as a reference the following values:

    Voice FM / SSTV / FM, 1200bps data: 15kHz

    Data 9600 or higher: 15kHz and broader in proportion to the baud rate

    SSB: 3kHz

    CW: 500Hz

Frequency coverage

It must obviously be able to receive on the downlink frequency of the satellite concerned. For the higher bands, it is common practice to extend the coverage through an external down converter (which will be discussed later on).

Demodulation

The receiver must obviously be able to demodulate the signals concerned, whether they are modulated in FM or SSB. This is a point to which careful attention should be paid when selecting the device: not all the commercial devices, especially the hand held or small size, are really "all-modes".

A feature that is gaining popularity is the ability to connect the receiver to a computer so that, for example, it is possible to automatically correct the tuning for compensating the Doppler Effect from your PC while you are engaged on other activities.

As for commercial examples that have had some success in recent years, we list the following receivers:

In recent years, the scenario is moving towards the SDR (Software Defined Radio) receiver, and will be undoubtedly be the future. With considerable operational advantage, today there are many different models on the market to choose from, some of which are not only bare receivers but also transmitters.

An SDR amateur project of great importance, and linked to the CubeSat developed by Amsat-UK is an SDR receiver integrated into a USB stick (**Fig 4.11**).

One of the main objectives of this project is to make a modern receiver of good performance available to many young students and telecommunications lovers.

More and updated information can be found on the following site:
www.funcubedongle.com/

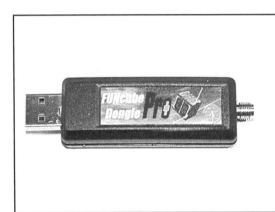 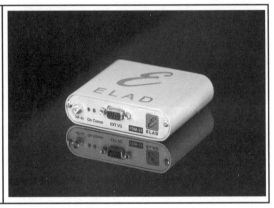

Fig 4.11: Fun Cube Dongle & Fun Cube Dongle Pro Plus   Fig 4.12: FDM-S1 Elad

Converter

All receivers and transceivers have an upper frequency limit that generally does not cover all the bands allocated to earth-space communications. To extend the coverage, converters are used which are divided into three main groups:

- Down-converter: Used for receiving where the input frequency is greater than the output: for example input 2400MHz and output 432MHz in order to listen to the Mode S with an UHF receiver.

- Up-converter: Used for transmission where the output frequency is higher than the input one: for example input 144MHz and output 1268MHz in order to uplink in L mode with a VHF transmitter.

- Transverter: Complex device that actually does both up and down conversion for full RX and TX capability. Its salient qualities are quite similar to those of pre-amplifiers for down-converters and the transmitter for up-converters. Particular attention should be paid to the accuracy and stability of the conversion oscillator.

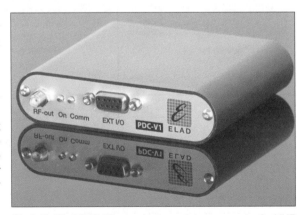

Fig 4.13: Elad PDC-V1 preselector and downconverter for VHF

The Ground Station

Here two more examples of commercial converters:

Fig 4.14: Down Converter 2400-144MHz model UEK-3000 from SSB Electronics

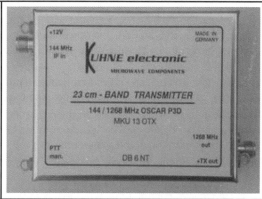

Fig 4.15: Up Converter 144-1268MHz, model MKU UP13 OTX from Kuhne

Transceiver

Many considerations made for simple receivers can be applied to transceivers.

Output power is not a matter of great importance: The value between 10W and 50W is suitable for the great majority of applications. Let's always remember that the satellite is substantially in the optical range and transmits to the ground with at most a few Watts (fractions of a Watt in the case of Cubesats!).

As for commercial transceivers particularly intended for our use which have had great success in recent years, we can remember:

ICOM IC-820	ICOM IC-910	ICOM IC-970
Kenwood TS-770	Kenwood TS-790	Kenwood TS-2000
Yaesu FT-726	Yaesu FT-736	Yaesu FT-847

Tracking system

The simplest tracking system consists of a table of time-position of the satellite and the operator's hand pointing the antenna accordingly. This system is very "spartan" and may give satisfaction with the LEO satellites, occasionally when working outdoors with small directional antennas.

In the case of a fixed station, it is obviously possible to have the same manual technique but it is often much more convenient to set up an automatic system tasked with tracking the satellite while we are busy with the radio.

A tracking system consists of two main parts: a program that calculates the current satellite position with respect to our station and an actuator pointing the antenna(s) coherently.

The software part of the satellite position calculation is perhaps the oldest one. There's already been software about for C64, Vic20, Spectrum, Amiga, etc. as early as twenty years ago, if not more, which allowed the calculation of the ephemeris.

Since then, much has changed and today's programs have excellent quality of accuracy, easy updating and various ancillary functions such as:

- Transceiver control: tuning, modes, compensation of the Doppler Effect
- Azimuth-elevation rotor control
- QSO logging

There are versions coming with GNU Licence, some of which are freeware while others are by payment for Licence. Program selection is very individual which involves personal taste and convenience as well as functionality needed for a specific use.

Let's see in the following table some of the most famous and popular tracking software now available:

Program name	OS	Radio control [3]	Rotor control [3]	Screenshot
Ham Radio DeLuxe 5.0 and above	Windows	yes	yes	
HamSatDroid	Android	no	no	

The Ground Station

Program name	OS	Radio control [3]	Rotor control [3]	Screenshot
Instant TRACK	DOS	no	no	
MacDoppler	OS10.4 e OS10.5	yes	yes	
NOVA for Windows	Windows	yes	yes [1,2]	
Orbitron	Windows	yes [1]	yes [1,2]	
PetitTrack	Embedded Linux	no	no	

49

Hamsat - Amateur Radio Satellites Explained

Program name	OS	Radio control [3]	Rotor control [3]	Screenshot
PocketSAT+	PalmOS, Win CE	no	no	
PocketSAT3	iOs	no	no	
Predict	OSx, Linux, SunOS, Windows	no	no	
SatExplorer	Windows	yes [1]	yes [1]	
SatPC32	Windows	yes	yes [1,2]	

Program name	OS	Radio control [3]	Rotor control [3]	Screenshot
SatScape	Windows	no	no	
Tracksat	HPC, WinCE, WinMobile	no	no	

Note:
1) with help of external program
2) with help of additional interface
3) check program documentation to see which models are supported

The second module is often provided by rotator suppliers although there are examples of "general purpose" interfaces suitable to operate with a plurality of commercial and home-made rotators as well as numerous solutions designed by the radio amateur.

Purely by way of example we can mention the following ones:

Web site	Compatible Rotors	Picture
www.ea4tx.com	**Yaesu**: G250, G400S, G450A, G500 & G500A (Ele), G650A, G800S, C, SDX & DXA, G1000S,C, SDX, DXA & DXC, G2000S & RC, G2800DXA, G5400, G5500, G5600, etc.; **Telex/Hy-Gain**: HAM-IV, HAM-M, HAM-2, T2X, HDR-300; **Kenpro**: KR-400RC, KR450XL, KR-500(Elevation), KR-600RC, KR.650XL, KR-600RC, KR600S, KR-800, KR-1000, KR-5400 & 5600 (Az&El); **Daiwa**: DR-7500 serie R and X, DR-7600 serie R and X; **ORION**: 2300; **CREATIVE**: RC5x-3P, RC5-1, RC5-x series, ERC-51; **EMOTATOR**: 1105, EV700X; **EMOTO**: 1200 FXX; **Aliance**: HD73 **Pro.Sis.Tel**: All "B" Control Unit models and "D" (**)	
www.funkbox.info	**Yaesu/Kenpro** G/KR 5400/5500/5600; **Emotator** EV 800 DX; **Create** RC/ERC; **Create** AER-5	
www.qsl.net/ve2dx/projects/fod.htm	Not available	
www.rrs-web.net/in3her/rotorsys_32.html	Not available	

The Ground Station

Web site	Compatible Rotors	Picture
www.hosenose.com/sar-tekx/	**Telex/Hy-Gain** Ham II, Ham III, Ham IV, Ham M, Tailtwister, CDX, CDE 45-II, HDR-300/A; **Alliance** HD-73; **Yaesu** G-800SDX, G-1000SDX, 2700SDX, 2800SDX, 800S, 1000S, 800DXA, 1000DXA, 2800DXA; Orion 2300; **Create** RC5A-2 & RC5A-3; **Imotator** 1300MSAX; **ProSisTel** tutti modelli B; **K0XG** rotating tower; **Kenpro** KR-2800SDX	

Audio Interfaces / PTT / CAT

When you want to connect the radio to a PC, it is often recommended and sometimes even necessary to use a specific interface primarily intended to electrically isolate the PC from the transceiver. In this way you can reduce the risk of ground loop and prevent disturbance and noise in the modulation. In addition to the electrical isolation of the audio signal, some of these devices have various functions such as the regulation of the level, the control of the PTT (Push to Talk), CW keying and also the CAT interface. The most recent radio models offer this feature by means of a USB port instead of the former serial RS232.

The schematic diagram of a simple and common interface is reproduced above **Fig 4.16**.

The galvanic separation between the ground of the sound card and the computer is achieved by means of two small audio transformers with unitary ratio and a photo coupler for the PTT line.

Lots more examples can be easily found in amateur radio magazines as well as on the internet.

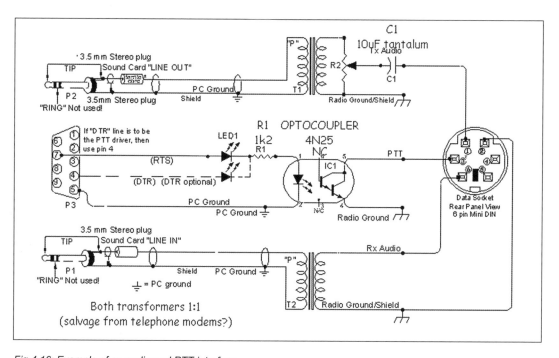

Fig 4.16: Example of an audio and PTT interface

Hamsat - Amateur Radio Satellites Explained

Below is a small gallery of amateur radio satellite stations, ranging from the most simple, inexpensive and portable to the most complex, sophisticated and expensive.

Call sign	Set-up	Picture
IW4BLG (2009)	Yaesu FT847; IBM A31p tracking, logging; Rotor interface: Funkbox; Yaesu FRG9600 as IF for S band	
I1TEX (2003)	Icom IC756; Home-made 2401MHz down converter; Kenwood TS790E for 435MHz (40W) and 1269MHz (250W)	
IW4BLG/p (2005)	Yaesu FT847; IBM T30; all that fit into the car	

The Ground Station

Call sign	Set-up	Picture
Politecnico Torino (2009)	Switchbox (transverter included: 28<->144 e 28<->432, switch RF, power splitters, power supplies, relays, sequencer, int. USB); IC-910H; TNC Multi-7; Perseus (RX-SDR); Control boxes SPID RAS; Radio power supply (Velleman PS1330); 2,4 GHz transverter power supply (Atten TPR-3010H); Rotors power supply (Atten TPR-3010H); PC	
IK1ODO (2010)	PC CPU: i3 with two monitors; Switchbox; IC910H; TNC7multi; Rotors control box; FT-736R; HF antenna distributor (Watkins-Johnson); Lowe HF-225 (RX HF); Tuner MFJ; Kenwood 599S; Rotor power supply; IC-910H power supply; VHF power amplifier (ITT AM-6155); Rubidium reference frequency generator 10MHz; ref. Frequency distributor; 30.2MHz synthesiser for Icom (in place of its internal TCXO); PTS 160; Spectrum analyzer - Tektronix 2754P – with spectrum surveillance SW; Signal generator - Fluke 6070A	

Call sign	Set-up	Picture
IV3CYF (2010)	ICOM IC 910H VHF/UHF; ICOM IC756 PRO 2 HF	

Reference
'A practical realisation of the notch may be the one proposed' by Heinz Hildebrand DL1CF in an old article

Ground Station Antenna

Antennas have a prominent role in the success of the amateur radio satellite communication, and so deserve a dedicated chapter.

What is the purpose of an antenna?

A generic antenna has the task of transforming the electrical input signal into an electromagnetic field and from electromagnetic field into an electric signal.

The electromagnetic wave

The physicist and mathematician James C Maxwell was the first to predict the existence of electromagnetic wave. As early as 1864 he had already proposed the wave theory of light and formulated the famous equations that are still the basis for any study of electromagnetism:

$$\nabla E = \frac{\rho}{\varepsilon_0}$$

$$\nabla B = 0$$

$$\nabla \times E = \frac{-\partial B}{\partial_t}$$

$$\nabla \times B = \mu_0 \dot{J} + \mu_0 \dot{\varepsilon}_0 \frac{\partial E}{\partial t}$$

In 1887 Heinrich Hertz provided the first experimental evidence of the existence of such a wave. Based on that research, the Italian scientist Guglielmo Marconi created what may be considered the first radio link in 1896.

In an electromagnetic wave, the magnetic field is generated by the displacement current caused by a varying electric field and, at the same time, the electric field is generated by the variation of the magnetic flux.

This intimate link between electric and magnetic field is the core of the existence of such a wave.

A plane wave is a particular type of wave whose descriptive parameters depend only on the time in one direction in space (eg the direction of propagation) and is uniform in all directions transverse to itself.

A key parameter: the G/T

For a long time, and as long as the transmission was mainly between a pair of ground stations, it used to be believed that the best antenna was the one with the higher gain, with the start of the space-to-ground traffic and the rising of the frequency involved, this assumption proved to be incomplete and often misleading.

It is not the maximum signal that provides the best connection but the best signal/noise ratio.

The numerator of this quotient is the signal, which is proportional to the antenna gain. While the denominator is the "disturbance", which represents, in this case, the "noise" that the antenna collects, the noise picked up by each antenna is generally the sum of three main contributions:
- noise equivalent to the loss of the antenna
- noise added by in the direction of the main lobe
- noise added by other directions

A small study on noise

Why so much attention to noise? Because noise is everywhere and it is often our biggest enemy! But what is noise?

One of the most famous definitions says:

> "Any stochastic function of time of specified length whose Fourier Series components each have a two dimensional normal distribution and random phase can be considered noise provided that the quadratic content of no single component is an appreciable percentage of the total".
> (EATON, Measuring noise)

It seems hard... although with a seemingly complex language, it expresses the concept of noise as a signal continuously variable, which has a very wide bandwidth and uniform energy content.

In a more general sense and relative to our amateur experience, noise is often defined as any source of "disturbance" superimposed on the useful signal that we want to receive.

There are many sources and kind of noise. One of the most interesting noises to study is the so-called "thermal noise" produced by our body as well as the stars.

Thermal noise is all around us: everything generates noise due to its physical temperature. This means, for example, that if you point the antenna at a tree with many leaves, you will receive the thermal noise generated by the foliage.

The thermal noise is a wideband signal and is measured in dBm/Hz. With an ambient temperature of 290K (17°C) the noise power is equal to -174dBm/Hz.

Thermal noise is very predictable and is determined by physical constants and therefore can be used in a simple manner as a reference. In fact, a way of expressing the noise level is using Kelvin and it is called noise temperature. The aforementioned -174dBm/Hz corresponds to a noise temperature of 290K.

The advantage of employing the noise temperature is to determine the total noise temperature of a multistage system with the contribution of each single stage added arithmetically. For example, if we have a noise temperature of the receiver of 75K, an antenna temperature of 50K and a loss of 0.1dB (= 6.8K) from the antenna to the receiver, the whole equivalent noise temperature will be 131.8K (50 +75 +6.8). Very quick and easy!

The bodies of space also produce noise called galactic noise generated both by their temperature (such as the Sun, the Moon, the Orion Nebula) or other complex mechanisms known as the "Synchrotron" (Cassiopeia A supernova remnant and the radio galaxy Cygnus A).

The galactic noise is well documented by radio astronomers and it can be used as a reference to evaluate the performance of our receiving station. Among the sources of extraterrestrial noise we can also mention the noise from the "big bang", the time of the birth of the universe. This noise level is very low (about 2.7K =-194dBm/Hz) and can be used as a reference of "low noise" which is also called "cold sky" for this reason.

Human activities are also a big source of noise and disturbance to our radio communications. The problem is particularly acute in urban areas but also increasingly present in rural area with the spreading of high speed data networks such as BPL, ADSL, switching power supplies (now present in almost all consumer electronics), and discharge lamps (neon, sodium, low power consumption, etc). The noise of human origin is not predictable and cannot be used to perform measurements on our station.

Other sources of noise are lightning and rain with its discharge of static electricity. They are intermittent and in most cases the interference is limited in time and at the lowest frequency of the radio spectrum.

Before going into the analysis, it should be noted that each antenna, even the best designed, will not only receive from the direction of the incoming signal, but also from other directions, albeit with reduced efficiency.

This characteristic is clearly visible in the radiation pattern as the example below:

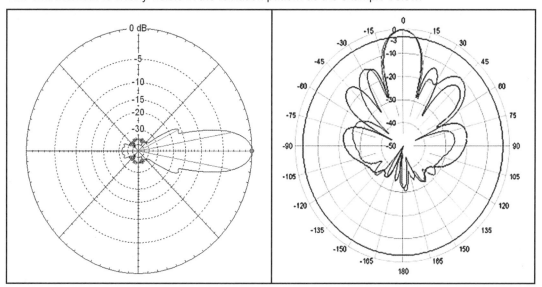

Fig 5.1: Pretty good antenna pattern   Fig 5.2: Poor antenna pattern

The preferred direction is called "primary lobe": the narrower it is, the higher the antenna gain and consequently the higher the signal received.

The sensitivity in the other directions is called "side lobes" and is an inevitable price to pay. Its value indicates how much noise (in the general meaning of "extraneous signal") the antenna picks up. Said noise level is generally converted into equivalent antenna temperature.

It is important to reiterate that the best reception is not the one in the presence of a strong signal but with the best signal/noise ratio, which in this particular situation is described by the ratio gain/noise temperature.

The noise temperature of a generic antenna is as follows:

$$T_{ant} = \frac{1}{4\Pi} \int\int_{00}^{2\Pi 2\Pi} t(x,y)$$

To calculate this value it is common practice to use specific software which takes the antenna pattern description and the equivalent noise temperature of the surrounding "world" as input. The program currently most widespread among radio amateurs is the TANT written by YT1NT capable of calculating antenna temperature and the ratio G/T for elevation between 0 and 90 degrees, in steps of 5 degrees by assuming the earth temperature and that of the sky which is uniform and constant in each hemisphere.

In the real world things are somewhat more complex, however this way can be a good starting point.

The amateur who has studied and promoted the knowledge of this subject more than anyone else is my good friend Lionel Edwards, VE7BQH.

He prepared the table below (for 144 and 432MHz), which shows a long list of antennas for 144MHz with the gain value of both single antenna and a group of four (typical configuration for EME) as well as the G/T.

144MHz antenna

TYPE OF ANTENNA	L (λ)	GAIN (dBd)	E (M)	H (M)	Ga (dBd)	Tlos (K)	Ta (K)	F/R (dB)	1st SL (dB)	2nd SL (dB)	Z (ohms)	VSWR Bandwidth	G/T (dB)	Feed System
+KF2YN Boxkite4	0.43	11.10	3.50	2.00	16.80	3.9	225.7	23.4	22.0	none	50.4	1.12:1	-4.59	Dipole
G4CQM 6	1.00	9.46	2.60	2.17	15.44	7.9	249.7	18.9	17.1	none	56.7	1.83:1	-6.38	Dipole
+KF2YN Boxkite 6	1.04	12.47	3.90	3.00	18.25	4.6	263.1	26.5	22.9	24.8	49.9	1.20:1	-3.80	Dipole
Vine 6 FD	1.10	9.69	2.64	2.21	15.67	8.2	238.4	24.1	18.4	none	48.3	1.18:1	-5.95	Folded Dipole
G0KSC 6LFA	1.13	9.69	2.60	2.19	15.64	4.0	236.9	24.5	19.8	none	49.3	1.04:1	-5.96	LFA Loop
DD0VF 6	1.16	9.73	2.63	2.22	15.71	5.5	240.1	23.7	16.4	none	27.2	1.07:1	-5.94	Dipole
*DD0VF 6	1.16	9.73	2.30	2.30	15.58	5.5	245.1	23.7	16.4	none	27.2	1.07:1	-6.16	Dipole
M2 2M7	1.28	9.94	2.65	2.26	15.76	3.7	245.0	18.4	16.1	none	204.9	1.14:1	-5.98	T Match
*M2 2M7	1.28	9.94	2.21	2.03	15.17	3.7	239.9	18.4	16.1	none	204.9	1.14:1	-6.48	T Match
+KF2YN Boxkite7	1.32	13.34	4.17	3.40	19.30	5.2	245.5	26.8	23.6	24.5	52.7	1.06:1	-2.40	Dipole
+YU7XL 8 Hybrid	1.34	10.50	2.79	2.50	16.40	3.2	251.6	19.8	17.1	none	199.9	1.13:1	-5.46	Horiz Dipole
*YU7XL 8 Hybrid	1.34	10.50	3.00	2.43	16.43	3.5	247.7	19.8	17.1	none	199.9	1.13:1	-5.36	Horiz Dipole
+G0KSC 7LFA	1.39	10.62	2.84	2.49	16.53	1.8	248.9	20.4	16.1	none	48.0	1.19:1	-5.28	LFA Loop
*G0KSC 7 LFA	1.39	10.62	2.60	2.20	16.20	1.8	233.6	20.4	16.1	none	48.0	1.19:1	-5.34	LFA Loop
+DG7YBN 7	1.44	10.59	2.88	2.47	16.55	4.5	242.7	23.2	17.4	none	47.2	1.70:1	-5.15	Bent Dipole
Vine 7 FD	1.45	10.56	2.83	2.46	16.47	8.2	238.6	22.8	17.9	none	47.9	1.14:1	-5.16	Folded Dipole
G4CQM 7	1.50	10.76	2.89	2.53	16.69	7.9	239.9	23.5	17.9	none	50.7	2.31:1	-4.96	Dipole
+CT1FFU 7	1.54	10.82	2.87	2.50	16.70	2.8	237.7	20.3	18.4	20.4	28.0	1.02:1	-4.96	
DK7ZB 7	1.57	11.11	3.16	2.84	17.13	5.8	272.6	16.9	11.9	16.9	28.4	1.64:1	-5.07	Dipole
IK0BZY 6	1.63	11.11	3.10	2.77	17.04	4.8	266.5	17.8	11.9	17.7	19.5	2.27:1	-5.07	
G4CQM 10 UZ2	1.67	10.74	2.89	2.51	16.68	5.7	235.9	25.3	19.5	25.3	45.1	1.26:1	-4.90	Dipole
+DG7YBN 8	1.68	10.94	2.91	2.56	16.84	3.5	238.8	19.7	15.5	20.2	47.5	1.16:1	-4.79	Dipole
G0KSC 8LFA	1.79	11.06	2.94	2.60	17.01	3.6	231.9	24.8	19.3	25.1	50.0	1.24:1	-4.49	LFA Loop
*G0KSC 8LFA	1.79	11.06	3.15	2.40	16.95	3.6	222.2	24.8	19.3	25.1	50.0	1.24:1	-4.37	LFA Loop
W1JR 8 MOD	1.80	11.14	3.07	2.75	16.99	5.3	256.7	17.3	12.8	17.3	50.0	1.14:1	-4.95	Dipole
DJ9BV 1.8	1.80	11.34	3.16	2.80	17.28	5.5	261.2	17.6	13.4	17.5	77.5	1.34:1	-4.74	Dipole
K1FO 10	1.84	11.34	3.10	2.78	17.27	4.3	257.7	16.5	14.5	18.5	29.4	1.44:1	-4.69	
Vine 8 FD	1.85	11.18	3.00	2.63	17.06	8.5	232.3	24.2	20.5	22.3	51.4	1.12:1	-4.45	Folded Dipole
YU7EF 8	1.87	11.31	3.04	2.71	17.23	3.8	242.1	20.0	15.1	20.1	48.5	1.21:1	-4.46	Dipole
BQH8B	1.88	11.60	3.28	2.97	17.62	7.2	259.3	18.0	12.6	16.5	50.0	1.29:1	-4.37	Dipole

Ground Station Antenna

TYPE OF ANTENNA	L (λ)	GAIN (dBd)	E (M)	H (M)	Ga (dBd)	Tlos (K)	Ta (K)	F/R (dB)	1st SL (dB)	2nd SL (dB)	Z (ohms)	VSWR Bandwidth	G/T (dB)	Feed System
+UR5EAZ 9	1.89	11.32	3.07	2.75	17.26	3.6	249.7	18.5	13.7	17.8	49.2	1.01:1	-4.56	Dipole
G4CQM 8	1.90	11.61	3.20	2.88	17.56	9.3	242.6	24.1	14.9	22.7	50.9	2.52:1	-4.14	Dipole
+KF2YN Box-kite9	1.92	13.98	4.45	3.70	19.95	5.6	228.6	24.4	21.5	26.3	49.2	1.28:2	-1.48	Dipole
+CT1FFU 8	1.94	11.28	2.96	2.62	17.10	2.9	232.3	23.6	21.2	21.6	27.1	1.05:1	-4.41	
G0KSC 8OWL	1.95	11.63	3.13	2.82	17.55	4.6	235.7	25.5	17.1	22.1	48.9	1.26:1	-4.02	Folded Dipole
I0JXX 8	2.04	12.11	3.46	3.17	18.10	9.3	257.3	19.6	13.2	17.5	200.1	3.00:1	-3.86	T Match
*DG0OPK 9	2.07	11.45	2.95	2.70	17.30	5.7	230.8	24.6	16.4	23.4	28.4	1.11:1	-4.18	Dipole
DG0OPK 9	2.07	11.45	3.04	2.72	17.34	5.7	231.9	24.6	16.4	23.4	28.4	1.11:1	-4.16	Dipole
DK7ZB 8	2.09	12.01	3.40	3.10	18.02	4.8	253.6	21.6	13.0	17.6	28.0	1.26:1	-3.87	Dipole
G0KSC 9OWA	2.09	11.99	3.33	3.04	17.96	4.9	247.0	21.3	15.0	17.1	49.1	1.30:1	-3.82	Dipole
+RA3AQ 9S	2.12	12.04	3.35	3.06	18.02	4.7	246.5	20.7	14.5	17.4	47.1	1.08:1	-3.75	Dipole
*RA3AQ 9S	2.12	12.04	3.00	3.00	17.85	4.6	242.8	20.7	14.5	17.4	47.1	1.08:1	-3.85	Dipole
M2 9SSB	2.12	11.96	3.33	3.04	17.92	10.8	245.9	20.2	14.7	17.1	200.6	1.26:1	-3.84	T Match
#WiMo WX220 XPOL	2.13	11.23	2.94	2.94	17.17	8.3	293.0	14.0	13.1	17.3	202.0	1.08:1	-5.35	Folded Dipole
*WiMo WX220 XPOL	2.13	11.23	2.94	2.94	17.44	8.4	300.8	14.0	13.1	17.3	202.0	1.08:1	-5.19	Folded Dipole
DJ9BV 2.1	2.13	11.89	3.33	3.04	17.87	5.8	255.2	20.2	13.2	17.4	44.9	1.30:1	-4.05	Dipole
G0KSC 9LFA	2.14	11.97	3.26	2.94	17.88	5.2	235.1	23.5	16.2	20.9	50.5	1.08:1	-3.68	LFA Loop
*G0KSC 9LFA	2.14	11.97	3.10	3.00	17.86	5.2	235.7	23.5	16.2	20.9	50.5	1.08:1	-3.75	LFA Loop
*OZ5HF 9	2.16	11.52	2.70	2.50	16.65	2.9	275.1	18.2	11.9	16.0	38.0	1.15:1	-5.60	
OZ5HF 9	2.16	11.52	3.21	2.92	17.47	3.0	272.6	18.2	11.9	16.0	38.0	1.15:1	-4.74	
YU7EF 9	2.16	11.84	3.20	2.89	17.74	5.3	236.6	22.0	16.7	19.8	49.8	1.17:1	-3.84	Dipole
F9FT 11	2.17	11.78	3.26	2.97	17.75	5.3	251.0	22.7	13.4	18.9	21.5	1.25:1	-4.10	
*CC 13B2	2.17	11.79	2.90	2.79	17.47	5.8	249.1	19.6	13.2	19.0	21.1	1.37:1	-4.34	
CC 13B2	2.17	11.79	3.33	3.01	17.78	5.9	257.9	19.6	13.2	19.0	21.1	1.37:1	-4.18	
K1FO 11	2.18	11.97	3.30	3.00	17.90	4.3	248.0	17.9	14.6	19.2	44.0	1.29:1	-3.89	
*CC 215WB	2.19	11.78	3.05	3.05	17.66	6.6	262.5	19.1	12.7	17.7	17.4	1.47:1	-4.38	
CC 215WB	2.19	11.78	3.36	3.07	17.78	6.6	266.5	19.1	12.7	17.7	17.4	1.47:1	-4.33	
Vine 9 FD	2.22	11.93	3.21	2.91	17.80	9.4	230.0	23.9	18.7	21.5	50.0	1.07:1	-3.67	Folded Dipole
+G0KSC 9OWL	2.28	12.15	3.28	2.97	18.00	6.4	229.5	24.3	18.8	22.3	49.0	1.20:1	-3.46	Folded Dipole
+KF2YN Boxkite 10	2.29	14.29	4.49	3.89	20.25	5.6	225.9	29.2	21.2	27.0	49.0	1.14:1	-1.13	Dipole
G4CQM CQM12UX	2.32	11.94	3.20	2.89	17.82	4.0	231.6	23.3	15.8	20.9	50.2	1.06:1	-3.68	Dipole
*Flexa 224	2.34	11.52	3.50	3.30	17.66	30.1	259.2	17.7	12.7	17.3	60.4	1.09:1	-4.33	
Flexa 224	2.34	11.52	3.30	3.00	17.47	29.5	254.6	17.7	12.7	17.3	60.4	1.09:1	-4.44	
+RA3AQ 9	2.35	12.38	3.40	3.13	18.31	5.6	234.3	22.1	15.1	18.8	49.2	1.10:1	-3.24	Folded Dipole
#RA3AQ 9	2.35	12.38	3.27	3.27	18.32	5.6	235.7	22.1	15.1	18.8	49.2	1.10:1	-3.25	Folded Dipole
+CT1FFU 9	2.38	12.23	3.26	2.97	18.08	3.6	226.8	22.7	20.2	21.2	28.1	1.15:1	-3.33	
ZL1RS 9	2.38	12.24	3.30	3.01	18.13	5.6	227.2	25.2	18.7	23.4	48.7	2.19:1	-3.28	Dipole
Eagle 10	2.38	12.28	3.44	3.16	18.27	6.0	243.0	22.0	15.0	19.2	23.6	1.33:1	-3.43	

Hamsat - Amateur Radio Satellites Explained

TYPE OF ANTENNA	L (λ)	GAIN (dBd)	E (M)	H (M)	Ga (dBd)	Tlos (K)	Ta (K)	F/R (dB)	1st SL (dB)	2nd SL (dB)	Z (ohms)	VSWR Bandwidth	G/T (dB)	Feed System
G4CQM CQM12UC	2.39	12.15	3.30	3.01	18.05	5.2	235.6	21.4	14.5	19.8	49.4	1.03:1	-3.54	Dipole
DK7ZB 9	2.39	12.41	3.56	3.30	18.45	6.2	250.5	20.4	12.2	16.0	27.5	1.23:1	-3.39	Dipole
Vine 10 FD	2.45	12.27	3.35	3.04	18.17	9.5	226.7	25.0	18.0	22.6	47.6	1.08:1	-3.23	Folded Dipole
DD0VF 9	2.46	12.47	3.48	3.18	18.44	4.4	235.8	20.3	15.8	17.5	25.0	1.16:1	-3.14	
+YU7EF 10LT	2.49	11.84	3.13	2.82	17.69	5.3	224.3	29.4	22.8	25.3	45.7	1.13:1	-3.67	Dipole
K5GW 10	2.49	12.45	3.44	3.16	18.38	7.4	234.6	23.1	16.4	21.4	37.4	1.41:1	-3.17	
#K5GW 10	2.49	12.45	3.30	3.30	18.38	7.3	235.9	37.4	16.4	21.4	23.1	1.41:1	-3.20	
G4CQM 9	2.52	12.63	3.62	3.35	18.68	7.3	244.4	22.2	13.1	17.6	46.4	2.02:1	-3.05	Dipole
+G0KSC 10 LFA	2.53	12.61	3.48	3.18	18.54	2.6	229.8	24.3	16.0	22.8	48.5	1.05:1	-2.92	LFA Loop
*G0KSC 10 LFA	2.53	12.61	3.40	3.10	18.48	2.5	228.3	24.3	16.0	22.8	48.5	1.05:1	-2.95	LFA Loop
K1FO 12	2.53	12.49	3.46	3.18	18.42	4.3	240.7	21.6	14.7	19.6	31.1	1.23:1	-3.25	
*YU7EF 10	2.59	12.44	3.38	3.09	18.33	6.9	227.5	23.8	17.2	23.1	50.3	1.06:1	-3.09	Dipole
YU7EF 10	2.59	12.44	3.39	3.10	18.34	6.9	227.8	23.8	17.2	23.1	50.3	1.06:1	-3.08	Dipole
+G0KSC 10LFA+2	2.60	12.45	3.39	3.09	18.35	4.1	226.7	23.8	18.9	22.6	47.4	1.24:1	-3.06	LFA Loop
'YU7XL 17 twin Bm	2.62	13.75	3.75	4.21	21.80	2.5	239.7	23.4	16.4	22.4	192.6	1.07:1	-2.02	Horiz Dipole
+G0KSC 10 OWL	2.63	12.46	3.40	3.10	18.36	7.2	224.6	26.4	19.1	26.2	48.2	1.11:1	-3.00	Folded Dipole
I0JXX 12	2.68	12.67	3.58	3.30	18.64	6.9	238.0	25.2	17.0	26.9	26.9	1.31:1	-2.97	
BQH 12J	2.80	12.80	3.67	3.40	18.83	4.3	246.2	20.3	13.5	17.2	50.0	1.01:1	-2.93	Dipole
#BQH 12J	2.80	12.80	3.54	3.54	18.81	4.3	248.0	20.3	13.5	17.2	50.0	1.01:1	-2.99	Dipole
+CT1FFU 10	2.82	12.74	3.44	3.15	18.81	4.1	220.2	25.2	20.0	23.2	28.1	1.19:1	-2.69	
+CT1FFU 10C	2.82	13.04	3.62	3.31	18.93	4.2	222.9	27.2	17.6	21.9	27.2	1.23:1	-2.40	
*M2 12	2.84	12.68	3.05	3.05	18.19	5.4	232.2	19.5	14.9	18.2	37.0	1.13:1	-3.32	
M2 12	2.84	12.68	3.48	3.20	18.54	5.4	232.5	19.5	14.9	18.2	37.0	1.13:1	-2.97	
+KF2YN Box-kite12	2.84	14.90	6.00	4.40	20.84	5.9	227.4	25.2	20.0	27.0	48.5	1.22:1	-0.58	Dipole
BQH 10	2.86	13.05	3.69	3.43	19.00	7.0	232.5	23.9	14.4	20.6	22.6	1.47:1	-2.51	
#BQH 10	2.86	13.05	3.56	3.56	19.00	6.9	232.8	23.9	14.4	20.6	22.6	1.47:1	-2.52	
WB9UWA 12	2.87	12.73	3.48	3.20	18.61	7.0	223.0	25.2	20.0	22.7	23.9	1.47:1	-2.72	
DK7ZB 10	2.87	13.00	3.87	3.60	19.17	6.8	247.5	21.7	12.7	16.6	25.8	1.61:1	-2.62	Dipole
Vine 11 FD	2.87	12.78	3.60	3.21	18.67	10.0	220.0	27.9	19.7	24.0	48.4	1.09:1	-2.60	Folded Dipole
+YU7EF 11B	2.87	12.92	3.58	3.30	18.85	4.7	225.6	25.6	16.7	20.1	50.2	1.35:1	-2.53	Dipole
I5MZY 13	2.88	13.21	3.82	3.56	19.21	7.8	238.1	22.8	15.1	17.8	57.2	2.45:1	-2.41	Dipole
+YU7XL 11 Hybrid	2.88	13.10	3.64	3.42	19.04	4.6	224.4	26.2	17.3	20.7	195.6	1.18:1	-2.32	Horiz Dipole
*YU7XL 11 Hybrid	2.88	13.10	3.77	3.56	19.08	4.8	226.7	26.2	17.3	20.7	195.6	1.18:1	-2.33	Horiz Dipole
K1FO 13	2.89	12.94	3.64	3.39	18.88	4.6	239.2	19.7	14.5	19.4	24.7	1.38:1	-2.76	
#M2 20 XPOL	2.97	13.11	3.64	3.64	19.09	6.3	241.7	21.4	13.5	16.1	200.1	1.20:1	-2.59	T Match
+G0KSC 11 LFA	2.98	13.11	3.61	3.35	19.01	3.0	221.5	27.7	16.5	24.8	48.3	1.06:1	-2.33	LFA Loop
*G0KSC 11 LFA	2.98	13.11	3.78	3.65	19.14	3.0	224.9	27.7	16.5	24.6	48.3	1.06:1	-2.23	LFA Loop
+UA9TC 11RS	2.98	13.09	3.66	3.37	19.01	5.0	223.0	23.3	17.9	21.5	50.9	1.10:1	-2.34	Dipole

Ground Station Antenna

TYPE OF ANTENNA	L (λ)	GAIN (dBd)	E (M)	H (M)	Ga (dBd)	Tlos (K)	Ta (K)	F/R (dB)	1st SL (dB)	2nd SL (dB)	Z (ohms)	VSWR Bandwidth	G/T (dB)	Feed System
+G0KSC 11OWL	3.00	13.16	3.69	3.44	19.09	7.7	224.6	23.0	16.4	22.3	49.8	1.42:1	-2.27	Folded Dipole
*BVO-3WL	3.01	13.43	3.90	3.70	19.38	8.6	253.1	20.5	12.9	17.7	52.7	3.08:1	-2.50	Dipole
BVO-3WL	3.01	13.43	4.03	3.78	19.42	8.6	256.1	20.5	12.9	17.7	52.7	3.08:1	-2.51	Dipole
#BVO-3WL	3.01	13.41	3.90	3.90	19.41	8.6	256.6	20.5	12.9	17.7	52.7	3.08:1	-2.53	Dipole
+G0KSC 11LFA3R	3.01	13.00	3.60	3.33	18.92	4.0	223.0	25.7	18.2	24.0	50.1	1.08:1	-2.41	LFA Loop
+YU7EF 11	3.04	13.07	3.56	3.30	18.87	4.7	222.6	22.2	17.8	22.5	49.0	1.62:1	-2.46	Dipole
F9FT 16	3.06	12.64	3.54	3.26	18.63	6.0	241.3	21.1	13.8	16.4	20.8	1.37:1	-3.05	
#SM2CEW 14 XPOL	3.08	13.04	3.50	3.50	18.96	6.9	226.4	25.8	22.5	25.9	19.2	1.40:1	-2.44	Dipole
CD15LQDver2	3.09	12.90	3.58	3.33	18.85	4.4	231.6	26.1	14.0	17.0	50.0	1.25:1	-2.65	Gamma Match
*CD15LQDver2	3.09	12.90	4.00	3.80	19.02	4.4	238.8	26.1	14.0	17.0	50.0	1.25:1	-2.61	Gamma Match
CD15LQDver1	3.10	12.83	3.60	3.35	18.76	4.1	247.8	20.7	14.5	17.5	50.0	1.23:1	-3.03	Gamma Match
*CD15LQDver1	3.10	12.83	4.00	3.80	18.97	4.2	249.9	20.7	14.5	17.5	50.0	1.23:1	-2.86	Gamma Match
I5MZY 13	3.10	12.97	3.56	3.30	18.83	6.0	225.9	20.3	15.1	17.8	49.3	1.50:1	-2.56	Dipole
MBI ModFT17	3.12	13.29	3.85	3.60	19.27	8.2	239.5	24.3	12.9	19.7	50.1	1.41:1	-2.37	Dipole
*F9FT 17	3.14	12.87	3.68	3.50	18.90	5.8	236.4	23.0	14.7	18.6	25.8	1.32:1	-2.69	
F9FT 17	3.14	12.87	3.59	3.31	18.81	5.7	234.3	23.0	14.7	18.6	25.8	1.32:1	-2.74	
*CC3219	3.17	12.77	4.27	3.66	18.80	5.7	307.8	14.9	15.1	18.0	18.3	1.49:1	-3.93	
CC3219	3.17	12.77	4.07	3.82	18.79	5.7	308.6	14.9	15.1	18.0	18.3	1.49:1	-3.95	
CC3219 MOD	3.17	13.25	3.87	3.62	19.27	5.1	245.6	24.1	13.0	16.0	29.0	1.06:1	-2.48	Dipole
BQH 13	3.19	13.30	3.85	3.60	19.28	4.3	244.9	20.8	13.7	17.5	50.0	1.11:1	-2.46	Dipole
#BQH 13	3.19	13.30	3.72	3.72	19.27	4.3	246.3	20.8	13.7	17.5	50.0	1.11:1	-2.49	Dipole
DJ9BV 3.2	3.22	13.30	3.85	3.58	19.29	6.5	239.5	21.1	13.7	18.1	71.8	1.36:1	-2.35	Dipole
+DG7YBN 12	3.23	13.45	3.87	3.62	19.41	5.5	229.1	26.2	15.0	20.1	46.2	1.40:1	-2.04	Bent Dipole
*DG7YBN 12	3.23	13.45	3.99	3.83	19.47	5.6	231.0	26.2	15.0	20.1	46.2	1.40:1	-2.02	Bent Dipole
K1FO 14	3.26	13.36	3.80	3.56	19.29	4.7	237.8	18.1	14.4	19.3	29.6	1.42:1	-2.32	
+KF2YN Boxkite 13	3.26	15.11	4.83	4.32	21.00	5.9	226.2	28.8	20.4	28.9	52.5	1.07:1	-0.38	Dipole
+G0KSC 12LFA	3.32	13.44	3.78	3.52	19.36	4.0	220.5	24.4	18.1	23.1	49.9	1.07:1	-1.92	LFA Loop
+G0KSC 12OWA	3.33	13.33	3.85	3.50	19.28	5.7	224.3	25.3	16.3	23.4	49.3	1.05:1	-2.08	LFA Loop
G4CQM 11	3.36	13.55	3.92	3.66	19.50	8.3	232.0	30.1	13.3	17.9	46.2	1.94:1	-2.01	Dipole
DK7ZB 11	3.40	13.68	3.94	3.71	19.61	5.0	234.3	22.3	14.2	17.4	27.9	1.27:1	-1.94	Dipole
+UA9TC 12RS	3.40	13.55	3.80	3.56	19.46	5.1	218.2	32.1	18.6	21.2	51.7	1.09:1	-1.78	Dipole
+G0KSC 12 LFA	3.41	13.64	4.59	3.60	19.62	4.3	224.1	26.1	18.1	23.1	50.1	1.18:1	-1.73	LFA Loop
*G0KSC 12LFA	3.41	13.64	3.90	3.70	19.58	4.3	221.2	26.1	18.1	23.1	50.1	1.18:1	-1.72	LFA Loop
MBI 3.4	3.42	13.58	3.87	3.62	19.49	9.2	227.7	23.0	16.0	19.9	37.7	1.62:1	-1.93	
+G0KSC 12LFA 2R	3.43	13.46	3.79	3.52	19.38	3.3	221.2	25.0	19.3	23.6	50.3	1.06:1	-2.11	LFA Loop
*G0KSC 12LFA 2R	3.43	13.46	3.95	3.75	19.48	4.9	220.5	25.0	19.3	21.6	50.3	1.06:1	-1.80	LFA Loop
InnoV 12 LFA	3.43	13.56	3.80	3.60	19.47	5.4	218.9	25.8	17.4	23.7	49.3	1.07:1	-1.78	LFA Loop

Hamsat - Amateur Radio Satellites Explained

TYPE OF ANTENNA	L (λ)	GAIN (dBd)	E (M)	H (M)	Ga (dBd)	Tlos (K)	Ta (K)	F/R (dB)	1st SL (dB)	2nd SL (dB)	Z (ohms)	VSWR Bandwidth	G/T (dB)	Feed System
YU7EF 12	3.49	13.67	3.85	3.60	19.55	6.1	221.1	23.6	16.8	21.9	45.4	1.77:1	-1.75	Dipole
*SM5BSZ 11	3.51	13.95	3.50	3.50	19.48	6.4	231.8	19.5	18.1	19.7	54.4	3.06:1	-2.02	Dipole
+SM5BSZ 11	3.51	13.95	4.05	3.80	19.81	6.3	238.8	19.5	18.1	19.7	54.4	3.06:1	-1.82	Dipole
*SM5BSZ 11A	3.53	14.01	4.00	4.00	19.92	6.1	245.8	16.7	16.6	16.8	52.3	3.03:1	-1.84	Dipole
+SM5BSZ 11A	3.53	14.01	4.13	3.92	19.93	6.1	245.4	16.7	16.6	16.8	52.3	3.03:1	-1.82	Dipole
17LQD EKM#1	3.59	13.30	3.75	3.52	19.21	4.2	242.1	20.8	15.1	18.1	50.0	1.23:1	-2.48	Gamma Match
17LQD EKM#2	3.59	13.39	3.75	3.50	19.32	4.5	227.5	25.6	15.3	19.6	50.0	1.25:1	-2.10	Gamma Match
+DL6WU 14	3.61	13.63	3.94	3.70	19.53	2.6	246.9	21.2	14.3	17.4	51.4	1.28:1	-2.24	Dipole
DJ9BV 3.6	3.61	13.67	4.00	3.78	19.57	5.4	249.7	21.3	13.9	18.0	53.5	1.30:1	-2.25	Dipole
K1FO 15	3.64	13.76	3.94	3.71	19.67	4.8	233.0	20.0	14.5	19.4	39.1	1.30:1	-1.85	
DK7ZB 12	3.83	14.17	4.21	4.00	20.11	6.9	237.4	25.6	13.6	17.9	26.8	1.46:1	-1.49	Dipole
+UA9TC 13RS	3.83	13.95	3.94	3.70	19.84	5.2	215.6	24.3	18.7	20.9	49.7	1.07:1	-1.35	Dipole
+G0KSC 13 LFA	3.84	14.10	4.07	3.85	20.02	4.4	221.1	27.4	16.5	23.7	48.5	1.14:1	-1.28	LFA Loop
+G0KSC 13 LFA	3.85	13.97	4.00	3.75	19.91	6.4	220.1	28.2	16.5	23.7	48.8	1.20:1	-1.37	LFA Loop
*G0KSC 13 LFA	3.85	13.97	4.06	3.90	19.93	6.4	220.5	28.2	16.5	23.7	48.8	1.20:1	-1.35	LFA Loop
InnoV 13 LFA	3.86	13.98	4.02	3.77	19.89	7.0	219.5	28.2	17.4	25.5	48.7	1.20:1	-1.37	LFA Loop
YU7EF 13M	3.86	13.85	3.92	3.66	19.73	7.0	218.8	25.7	18.6	21.1	49.9	1.05:1	-1.52	Dipole
+YU7EF 13	3.92	14.12	4.05	3.82	20.00	6.0	220.1	22.5	17.0	23.2	47.9	2.55:1	-1.28	Dipole
IK0BZY 12	3.95	14.06	4.02	3.78	19.93	6.9	222.5	24.6	18.6	20.7	99.2	1.44:1	-1.39	Folded Dipole
+DG7YBN 14	3.98	14.09	4.05	3.82	20.00	5.8	218.8	28.3	17.4	20.4	50.4	1.07:1	-1.25	Bent Dipole
*DG7YBN 14	3.98	14.09	4.20	4.00	20.08	5.8	220.4	28.3	17.4	20.4	50.4	1.07:1	-1.20	Bent Dipole
BVO2-4WL	3.99	14.17	4.24	4.02	20.14	6.8	241.8	24.0	13.3	17.5	45.0	1.25:1	-1.54	Dipole
#BVO2-4WL	3.99	14.17	4.13	4.13	20.11	6.8	241.2	24.0	13.3	17.5	45.0	1.25:1	-1.56	Dipole
DJ9BV 4.0	4.01	14.04	4.16	3.92	19.91	6.5	249.4	22.4	14.3	18.3	41.5	1.35:1	-1.91	Dipole
K1FO 16	4.01	14.13	4.16	3.87	20.05	4.7	229.9	23.9	14.7	19.8	35.0	1.15:1	-1.42	
+SV 2SA13	4.01	14.45	4.38	4.16	20.42	7.0	239.0	20.0	14.2	17.3	52.2	1.41:1	-1.21	Dipole
#SV 2SA13	4.01	14.45	4.27	4.27	20.42	7.0	238.8	20.0	14.2	17.3	52.2	1.41:1	-1.21	Dipole
HG VB-215DX	4.03	14.10	4.21	3.97	20.00	5.8	250.6	19.6	14.6	18.6	38.6	1.41:1	-1.84	
CC3219 MOD	4.04	14.13	4.27	4.07	20.09	5.1	248.2	24.0	12.2	17.6	32.1	1.02:1	-1.71	Dipole
KLM 16LBX	4.09	14.13	4.26	4.02	20.11	6.2	241.3	20.2	14.9	18.7	198.1	1.29:1	-1.58	Dual Driven
*CC4218XL	4.19	13.92	4.08	3.85	19.74	11.1	260.2	16.8	14.3	19.2	14.5	1.72:1	-2.26	
CC4218XL	4.19	13.92	4.31	4.07	19.82	11.1	263.6	16.8	14.3	19.2	14.5	1.72:1	-2.24	
CC4218 MOD	4.17	14.27	4.24	4.02	20.21	5.8	236.9	22.0	13.8	18.2	22.0	1.18:1	-1.39	
WB9UWA 15	4.18	13.65	3.75	3.50	19.44	8.3	208.9	34.0	25.3	28.8	26.0	1.51:1	-1.61	
+G0KSC 14LFA3R	4.23	14.44	4.19	3.97	20.35	2.4	216.5	27.7	16.9	23.7	48.8	1.07:1	-0.86	LFA Loop
*G0KSC 14LFA3R	4.23	14.44	4.19	4.00	20.35	2.4	216.5	27.7	16.9	23.7	48.8	1.07:1	-0.86	LFA Loop
YU7EF 14M	4.24	14.23	4.07	3.85	20.08	7.0	220.1	24.6	18.6	20.9	48.6	1.07:1	-1.20	Dipole
+UA9TC 14RS	4.24	14.33	4.13	3.80	20.22	5.2	216.7	25.0	18.5	20.3	49.5	1.06:1	-0.99	Dipole
+G0KSC 14 LFA	4.30	14.51	4.24	4.02	20.41	5.5	218.1	26.8	16.9	24.5	47.1	1.22:1	-0.82	LFA Loop
*G0KSC 14 LFA	4.30	14.51	4.16	3.94	20.37	5.5	217.1	26.8	16.9	24.5	47.1	1.22:1	-0.84	LFA Loop

Ground Station Antenna

TYPE OF ANTENNA	L (λ)	GAIN (dBd)	E (M)	H (M)	Ga (dBd)	Tlos (K)	Ta (K)	F/R (dB)	1st SL (dB)	2nd SL (dB)	Z (ohms)	VSWR Bandwidth	G/T (dB)	Feed System
+KF2YN Boxkite 16	4.30	15.85	5.50	5.10	21.83	6.2	227.4	29.4	19.1	25.0	50.5	1.16:1	0.41	Dipole
InnoV 14 LFA	4.32	14.45	4.18	3.97	20.34	6.6	215.6	27.6	17.3	25.2	48.6	1.10:1	-0.85	LFA Loop
*InnoV 14 LFA	4.32	14.45	4.00	4.00	20.30	214.8	214.8	27.6	17.3	25.2	48.6	1.10:1	-0.87	LFA Loop
YU7EF 14	4.37	14.48	4.21	4.00	20.33	8.8	220.0	22.8	17.5	22.8	41.0	2.30:1	-0.94	Dipole
YU7EF 14LT	4.37	13.91	3.89	3.64	19.60	11.7	211.0	29.1	23.2	24.7	47.8	1.29:1	-1.49	Dipole
K1FO 17	4.40	14.44	4.24	4.02	20.34	4.9	229.5	22.9	14.7	19.6	27.7	1.25:1	-1.12	
DJ9BV 4.4	4.41	14.32	4.27	4.07	20.18	6.6	251.5	19.6	14.5	18.1	38.3	1.36:1	-1.68	Dipole
SHARK 20	4.47	14.34	4.33	4.10	20.19	3.9	258.5	22.4	13.3	18.0	42.8	1.05:1	-1.79	Dipole
I0JXX 16	4.47	14.36	4.16	3.94	20.24	8.4	217.6	27.6	17.9	22.2	30.4	1.05:1	-0.99	
#I0JXX 16	4.47	14.36	4.05	4.05	20.24	8.4	217.5	27.6	17.9	22.2	30.4	1.05:1	-0.98	
*CC17B2	4.49	14.52	3.66	3.51	19.93	6.1	223.4	23.4	15.6	21.6	29.3	1.22:1	-1.41	
CC17B2	4.49	14.52	4.27	4.07	20.42	6.2	226.8	23.4	15.6	21.6	29.3	1.22:1	-0.99	
RA3AQ-14	4.61	14.71	4.27	4.07	20.55	5.0	219.1	28.4	15.8	20.5	50.0	1.19:1	-0.71	Folded Dipole
G4CQM 16	4.64	14.55	4.39	4.18	20.47	8.8	234.6	34.0	12.8	17.0	50.9	1.78:1	-1.08	Dipole
YU7EF 15M	4.68	14.59	4.21	4.00	20.40	7.0	221.7	25.7	18.1	20.8	48.6	1.03:1	-0.89	Dipole
DK7ZB 14	4.73	14.92	4.59	4.39	20.87	6.8	233.0	26.2	13.0	18.2	26.9	1.50:1	-0.65	Dipole
+DG7YBN 16	4.74	14.79	4.39	4.21	20.70	6.1	221.4	28.9	15.8	20.6	46.6	1.18:1	-0.60	Bent Dipole
*DG7YBN 16	4.74	14.79	4.45	4.60	20.79	6.1	222.4	28.9	15.8	20.6	46.6	1.18:1	-0.53	Bent Dipole
G0KSC 15LFA	4.75	14.72	4.27	4.07	20.59	5.9	213.2	30.3	19.6	24.5	49.9	1.17:1	-0.55	LFA Loop
*G0KSC 15LFA	4.75	14.72	4.60	4.45	20.72	5.9	215.0	30.3	19.6	24.5	49.9	1.17:1	-0.45	LFA Loop
InnoV 15 LFA	4.76	14.73	4.30	4.10	20.62	6.6	213.9	29.8	17.1	26.3	49.0	1.17:1	-0.53	LFA Loop
*InnoV 15 LFA	4.76	14.73	4.60	4.45	20.74	6.6	215.4	29.8	17.1	26.3	49.0	1.17:1	-0.44	LFA Loop
K1FO 18	4.78	14.37	4.36	4.16	20.62	2.8	228.6	20.5	14.8	19.1	199.9	1.32:1	-0.82	T Match
InnoV 15 OWL	4.78	14.79	4.36	4.15	20.66	7.6	217.9	23.8	15.9	23.1	52.2	1.32:1	-0.57	Folded Dipole
*M2 28 XPOL	4.81	15.18	4.50	4.50	20.98	13.9	244.8	20.2	13.4	18.9	200.1	5.19:1	-0.76	T Match
#M2 28 XPOL	4.81	15.18	4.83	4.83	21.14	13.9	243.1	20.2	13.4	18.9	200.1	5.19:1	-0.57	T Match
DJ9BV 4.8	4.82	14.63	4.40	4.19	20.48	6.6	242.8	21.0	14.7	18.0	49.7	1.29:1	-1.22	Dipole
*M2 5WL	4.83	14.64	4.15	3.84	20.49	7.2	246.3	20.0	14.3	17.6	200.9	1.42:1	-1.44	T Match
M2 5WL	4.83	14.64	4.56	4.36	20.58	3.7	248.1	20.0	14.3	17.6	200.9	1.42:1	-1.22	T Match
YU7EF 15	4.84	14.89	4.46	4.24	20.78	10.3	218.1	25.6	17.9	22.8	41.8	3.18:1	-0.46	Dipole
+RU1AA_2	4.89	15.06	4.62	4.42	20.98	6.6	232.5	24.3	15.0	17.0	50.0	1.25:1	-0.53	Dipole
*SM5BSZ 14A	4.89	15.13	4.00	4.00	20.65	7.8	231.5	20.7	18.6	20.3	55.3	5.63:1	-0.85	Dipole
+RA3AQ 15	4.92	15.16	4.67	4.49	21.11	6.8	232.4	24.4	14.2	17.6	51.7	1.12:1	-0.40	Dipole
#RA3AQ 15	4.92	15.16	4.59	4.59	21.11	6.8	231.9	24.4	14.2	17.6	51.7	1.12:1	-0.39	Dipole
*SM5BSZ 14	4.95	15.27	5.20	5.20	21.36	5.7	239.7	18.8	14.9	19.1	49.3	4.37:1	-0.29	Dipole
+SM5BSZ 14	4.95	15.27	4.73	4.52	21.14	5.7	241.5	18.8	14.9	19.1	49.3	4.37:1	-0.54	Dipole
K5GW 17	4.98	14.83	4.43	4.21	20.73	9.8	220.8	25.7	15.9	20.5	48.5	1.73:1	-0.56	Dipole
SM2CEW 19	4.99	14.92	4.49	4.27	20.83	9.0	223.4	22.1	15.3	20.2	78.4	1.62:1	-0.51	Folded Dipole
#SM2CEW 19	4.99	14.92	4.38	4.38	20.83	9.0	223.1	22.1	15.3	20.2	78.4	1.62:1	-0.50	Folded Dipole
+G0KSC 16 OWL FD	4.99	14.75	4.25	4.25	20.63	9.0	218.8	28.4	19.2	25.5	51.7	1.10:1	-0.62	Folded Dipole

Hamsat - Amateur Radio Satellites Explained

TYPE OF ANTENNA	L (λ)	GAIN (dBd)	E (M)	H (M)	Ga (dBd)	Tlos (K)	Ta (K)	F/R (dB)	1st SL (dB)	2nd SL (dB)	Z (ohms)	VSWR Bandwidth	G/T (dB)	Feed System
*BVO-5WL	5.02	15.00	4.58	4.40	20.92	6.9	237.2	26.3	13.3	17.1	47.4	1.26:1	-0.68	Dipole
#BVO-5WL	5.02	15.00	4.58	4.58	20.97	6.9	236.2	26.3	13.3	17.1	47.4	1.26:1	-0.61	Dipole
BVO-5WL	5.02	15.00	4.66	4.49	20.96	6.9	236.9	26.3	13.3	17.1	47.4	1.26:1	-0.64	Dipole
YU7EF 16M	5.12	14.90	4.36	4.16	20.70	7.0	221.9	25.1	17.9	21.3	49.9	1.03:1	-0.61	Dipole
+G0KSC 16LFA3R	5.14	15.12	4.46	4.24	20.94	4.6	213.7	25.3	17.5	23.7	49.9	1.10:1	-0.21	LFA Loop
+G0KSC 16LFA3R	5.14	15.12	5.10	5.10	21.16	4.8	212.0	25.3	17.5	23.7	49.9	1.10:1	0.05	LFA Loop
K1FO 19	5.16	15.10	4.49	4.27	20.98	3.4	225.4	21.0	18.8	18.8	204.9	1.25:1	-0.40	T Match
+G0KSC 16LFA	5.21	15.40	4.56	4.33	21.29	11.0	212.7	28.5	17.5	24.4	49.2	1.13:1	0.16	LFA Loop
*G0KSC 16LFA	5.21	15.40	4.60	4.40	21.31	11.0	212.8	28.5	17.5	24.4	49.2	1.13:1	0.18	LFA Loop
InnoV 16 LFA	5.23	15.12	4.52	4.23	20.99	7.0	213.7	29.9	17.7	25.5	47.3	1.13:1	-0.16	LFA Loop
#RU1AA 15	5.27	15.44	4.78	4.78	21.38	9.2	230.9	25.2	13.9	18.6	54.6	3.70:1	-0.10	Dipole
RU1AA 15	5.27	15.44	4.87	4.70	21.37	9.2	231.7	25.2	13.9	18.6	54.6	3.70:1	-0.13	Dipole
*M2 18XXX	5.30	15.00	4.27	3.96	20.54	7.2	232.8	23.1	16.4	19.6	199.7	1.07:1	-0.98	T Match
M2 18XXX	5.30	15.00	4.56	4.36	20.90	7.4	231.9	23.1	16.4	19.6	199.7	1.07:1	-0.60	T Match
YU7EF 16	5.42	15.15	4.46	4.27	20.95	7.3	218.9	22.1	18.4	24.3	43.9	2.50:1	-0.30	Dipole
G0KSC 17 LFA	5.67	15.43	4.63	4.43	21.30	4.9	212.0	30.8	17.6	26.1	48.3	1.23:1	0.19	LFA Loop
*G0KSC 17 LFA	5.67	15.43	4.70	4.47	21.32	4.9	212.1	30.8	17.6	26.1	48.3	1.23:1	0.20	LFA Loop
InnoV 17 LFA	5.69	15.36	4.59	4.39	21.22	7.2	211.8	32.0	17.6	28.6	50.0	1.26:1	0.11	LFA Loop
*M2 19XXX	5.71	15.28	4.27	4.04	20.82	7.5	235.8	24.5	15.9	16.2	201.1	1.42:1	-0.75	T match
M2 19XXX	5.71	15.28	4.70	4.52	21.20	7.7	231.3	24.5	15.9	16.2	201.1	1.42:1	-0.29	T match
#M2 32 XPOL	5.74	15.81	5.08	5.08	21.77	12.8	232.4	23.2	13.4	18.3	205.4	3.80:1	0.26	T Match
*M2 32 XPOL	5.74	15.81	5.00	5.00	21.74	12.8	232.8	22.2	13.4	18.3	205.4	3.80:1	0.22	T Match
YU7EF 17X	5.74	15.36	4.59	4.39	21.19	8.5	216.0	29.2	20.0	22.0	50.0	1.08:1	-0.02	Dipole
YU7EF 17X	5.74	15.36	4.50	4.35	21.15	8.5	215.8	29.2	20.0	22.0	50.0	1.08:1	-0.04	Dipole
RU1AA 17	5.75	15.64	4.81	5.00	21.55	8.1	229.6	25.9	14.7	16.7	50.0	1.08:1	0.09	Dipole
+G0KSC 17OWL-FD	5.77	15.62	5.15	5.00	21.63	7.3	221.1	22.9	18.1	20.9	112.0	1.10:1	0.33	Folded Dipole
DK7ZB 17	5.82	15.61	5.01	4.85	21.59	8.6	227.8	24.8	12.7	17.7	26.8	1.19:1	0.16	Dipole
YU7EF 17	5.88	15.70	4.85	4.63	21.53	9.6	223.8	24.0	18.7	21.3	41.2	2.92:1	0.18	Dipole
#YU7EF 17	5.88	15.70	4.74	4.74	21.53	9.6	223.6	24.0	18.7	21.3	41.2	2.92:1	0.19	Dipole
BVO-6WL	6.00	15.65	4.93	4.77	21.55	7.0	227.3	24.9	14.3	18.2	44.9	1.14:1	0.13	Dipole
#BVO-6WL	6.00	15.65	4.85	4.85	21.59	7.1	226.7	24.9	14.3	18.2	44.9	1.14:1	0.19	Dipole
+G0KSC 18 LFA	6.12	15.69	4.77	4.56	21.55	5.3	210.5	30.8	17.7	25.4	47.9	1.33:1	0.47	LFA Loop
*G0KSC 18 LFA	6.12	15.69	4.77	4.60	21.56	5.3	210.5	30.8	17.7	25.4	47.9	1.33:1	0.48	LFA Loop
InnoV 18 LFA	6.14	15.66	4.73	4.56	21.52	7.7	211.0	32.7	17.6	25.3	48.7	1.35:1	0.43	LFA Loop
AF9Y 22	6.14	15.73	5.01	4.85	21.66	13.1	221.4	24.9	12.2	18.0	49.9	2.65:1	0.36	Folded Dipole
+RA3AQ 18	6.28	16.09	5.13	4.97	22.01	8.0	223.4	26.9	15.2	19.6	54.9	1.13:1	0.67	Dipole
*RA3AQ 18	6.28	16.09	5.30	5.30	22.12	8.1	222.3	26.9	15.2	19.6	54.9	1.13:1	0.80	Dipole
#RA3AQ 18	6.28	16.09	5.05	5.05	22.01	8.0	222.9	26.9	15.2	19.6	54.9	1.13:1	0.68	Dipole
MBI 6.6	6.58	16.15	5.50	5.31	22.17	11.5	228.3	26.9	12.4	18.9	49.1	1.77:1	0.73	Folded Dipole
#MBI 6.6	6.58	16.15	5.41	5.41	22.17	11.5	228.2	26.9	12.4	18.0	49.1	1.77:1	0.74	Folded Dipole

Ground Station Antenna

TYPE OF ANTENNA	L (λ)	GAIN (dBd)	E (M)	H (M)	Ga (dBd)	Tlos (K)	Ta (K)	F/R (dB)	1st SL (dB)	2nd SL (dB)	Z (ohms)	VSWR Bandwidth	G/T (dB)	Feed System
DK7ZB 19	6.59	16.15	5.41	5.22	22.15	8.2	230.2	24.0	13.4	17.9	27.7	1.97:1	0.68	Dipole
InnoV 19 LFA	6.62	15.88	4.84	4.70	21.74	8.0	209.3	32.5	17.5	26.9	48.3	1.55:1	0.68	LFA Loop
*InnoV 19 LFA	6.62	15.88	4.84	4.75	21.75	8.0	209.2	32.5	17.5	26.9	48.3	1.55:1	0.69	LFA Loop
BQH 25	7.30	16.38	5.27	5.09	22.28	8.8	217.1	25.5	15.2	20.8	27.9	1.16:1	1.06	Dipole
#BQH 25	7.30	16.38	5.18	5.18	22.28	8.9	216.8	25.5	15.2	20.8	27.9	1.16:1	1.07	Dipole
InnoV 21 LFA	7.55	16.25	5.04	4.88	22.10	8.2	208.3	32.5	19.0	26.6	49.0	1.70:1	1.06	LFA Loop
K2GAL 21	7.65	16.90	5.71	5.55	22.81	13.3	223.9	26.3	15.7	19.1	17.2	9.01:1	1.46	Dipole
M2 8WL(old)	7.72	16.51	5.31	5.13	22.34	9.1	221.9	21.3	16.3	19.1	201.5	2.14:1	1.03	T Match
InnoV 22 LFA	8.01	16.39	5.09	4.88	22.19	7.7	208.0	33.5	19.6	30.3	48.6	1.38:1	1.16	LFA Loop
M2 8WLHLD	8.05	17.02	6.05	5.87	22.99	11.8	234.6	25.4	13.2	19.4	200.0	4.53:1	1.44	T Match
+G0KSC 22 LFA 3R	8.17	16.37	6.10	5.90	22.44	6.7	208.1	29.6	20.6	29.2	46.1	1.16:1	1.43	LFA Loop
+G0KSC 22 LFA 3R	8.17	16.37	5.90	5.70	22.42	6.8	207.6	29.6	20.6	29.2	46.1	1.16:1	1.42	LFA Loop

Legend:
1. L = Length of wavelength
2. Gain = Gain in dBd of a single antenna
3. E = E plane (Horizontal) stacking in meters.
4. H = H plane (Vertical) stacking in meters.
5. Ga = Gain in dBd of a 4 bay array
6. Tlos = Internal resistance of the antenna in degrees Kelvin.
7. Ta = Total temperature of the antenna or array in degrees Kelvin which includes all the side lobes, rear lobes and internal resistance of the antenna or array.
8. F/R = Front to Rear in dB over the rear 180 degree of an antenna using either E or H plane.
9. Z Ohms = The natural impedance of a single antenna in free space.
10. VSWR = VSWR Bandwidth is based a single antenna over 144.000 - 145.000MHz with a reference at 144.100MHz. This parameter gives an indicator of the antenna "Q" and what to expect with stacking and wet weather.
11. G/T = Figure of merit used to determine the receiving capability of the antenna or array = (Ga + 2.15) - (10*log Ta). The more positive figure, the better. G/T is modeled in Tant.exe at 30 degrees elevation.

Notes:
1. The programs used to calculate E/H Stacking, G, Ta, Tlos and G/T are EZNEC 5+ by Roy Lewallen[1] W7EL, 4NEC2 by Arie Voors and Tant.exe by Sinisa, YT1NT/VE3EA [2]. This combination of software provides excellent accuracy. Segment Density is 25 segments per half wave.
2. Temperatures used: Tsky=200 degree; Tearth=1000 degree
3. Dipole Z is measured at 144.1MHz
4. F/R, 1st and 2nd Side Lobes (SL) have been calculated in a single antenna
5. No stacking harness loss or H frame effect are included in the 4 bay gain figures.
6. All stacking dimensions EXCEPT those marked with a "*" and "#" are calculated from the DL6WU stacking formula:
   $D = W/(2* \sin(B/2))$

Where:

- D = stacking distance, vertical or horizontal
- W = wavelength, in the same unit as D
- B – beam-width between -3dB points.
- Use vertical beam-width for vertical stacking (as above),
- Use horizontal beam-width for horizontal stacking.

7. Antennas marked with a "*" have stacking dimensions recommended by the manufacturer or designer.
8. Antennas marked with a "#" have stacking dimensions for XPOL antennas by VE7BQH.
9. Antennas marked with a "+" have some or all elements over 6mm. All others are 4MM to 6MM.
10. FD = Folded Dipole
11. Manufacturer/Designer Legend:
12. Convergence Correction: NEC2 and NEC 4 are incapable of handling complex feed systems accurately like Folded Dipoles, T Matches, LFAs etc. Convergence Correction using the KF2YN Excel program is required. See DUBUS 4/2010 "The Correction of Convergence Errors in Antenna Temperature Calculations by Brian Cake, KF2YN for details.
13. Manufacturer/Designer Legend:

AF9Y = AF9Y	WiMo = WiMo
BVO = Eagle/DJ9BV	K1FO = K1FO
BQH = VE7BQH	K2GAL = K2GAL
CC = Cushcraft	K5GW = Texas Towers/K5GW
CC MOD = VE7BQH	KF2YN = KF2YN
CD = CUE DEE	M2 = M²
CD MOD = VE7BQH	MBI = F/G8MBI/F5VHX
CT1FFU = CT1FFU	OZ5HF = Vargarda
DD0VF = DD0VF	RA3AQ = RA3AQ
DJ9BV = DJ9BV	RU1AA = RU1AA
DJ9BV OPT = DJ9BV	SHARK = SHARK (Italian)
DK7ZB = DK7ZB	SM2CEW = SM2CEW/VE7BQH
EKM MOD = SM2EKM	SV = Svenska Antennspecialisten AB
F9FT = F9FT	W1JR = VE7BQH (Mininec error)
Flexa = FlexaYagi	WB9UWA = WB9UWA
G0KSC LFA = G0KSC	YU7EF = YU7EF
G4CQM = G4CQM	UR5CSZ = UR5CSZ
HG = HYGAIN	UA9TC = UA9TC
I0JXX = I0JXX	Vine = G0KSC Design
IK0BZY = IK0BZY	YU7XL = YU7XL

Using this Chart

While Gain is important, other factors like easy matching and wet weather performance should be considered in the decision making. Antennas with 50 Ohm feed system and good VSWR bandwidth (Q) may be the best choice depending on your location. Low side lobe and F/R antennas with good (G/T) may provide further significant benefit if you have local man-made noise that is in the direction where these kinds of antennas provide additional suppression.

Lionel H. Edwards

VE7BQH

Issue 91, December 14, 2012

432MHz Antenna

TYPE OF ANTENNA	L λ	GAIN (dBd)	E (M)	H (M)	Ga (dBd)	Tlos (K)	Ta (K)	F/R (dB)	1st SL (dB)	2nd SL (dB)	Z (ohms)	VSWR Bandwidth	G/T (dB)	Feed System
KF2YN Boxkite 7	1.34	15.64	1.40	1.17	21.56	8.1	42.3	27.6	27.5	21.3	51.7	1.05:1	5.31	Bent Dipole
KF2YN Boxkite 10	2.37	16.55	1.50	1.30	22.46	8.7	38.3	28.3	19.2	28.0	48.7	1.08:1	6.64	Bent Dipole
InnoV 10 LFA	2.43	14.43	1.12	1.03	20.36	3.7	36.4	23.8	17.2	23.3	50.7	1.13:1	4.76	LFA
YU7EF EF7010-5	2.59	14.64	1.14	1.05	20.56	5.0	37.0	22.5	16.9	24.1	50.0	1.06:1	4.88	Dipole
YU7EF EF7011B-5	2.87	15.07	1.20	1.11	21.00	4.7	36.1	25.4	16.9	20.1	49.0	1.56:1	5.43	Dipole
Innov 11 LFA	2.87	15.05	1.20	1.10	20.98	4.0	33.7	25.6	16.4	23.8	50.4	1.07:1	5.70	LFA
*Innov 11 LFA	2.87	15.05	1.19	1.10	20.96	4.0	33.6	25.6	16.4	23.8	50.4	1.07:1	5.70	LFA
*KF2YN Polly 12 CR	3.17	15.70	1.30	1.30	21.74	3.19	32.27	25.4	16.1	22.1	48.3	1.17:1	6.65	Multi-Pol loop
InnoV 12 LFA	3.29	15.48	1.24	1.16	21.40	4.0	31.2	28.3	16.8	23.3	50.1	1.07:1	6.46	LFA
*InnoV 12 LFA	3.29	15.48	1.23	1.15	21.37	4.0	31.0	28.3	16.8	23.3	50.1	1.07:1	6.46	LFA
YU7EF EF7012B-5	3.30	15.58	1.26	1.17	21.50	4.9	33.3	24.0	17.6	21.3	48.6	1.40:1	6.27	Dipole
KF2YN Boxkite 13	3.37	17.38	1.63	1.50	23.29	9.3	39.4	28.0	18.0	29.0	50.9	1.09:1	7.35	Bent Dipole
Innov 13 LFA	3.72	15.92	1.31	1.23	21.85	4.5	30.9	29.5	16.5	23.2	50.1	1.15:1	6.96	LFA
*innoV 13 LFA	3.72	15.92	1.30	1.22	21.83	4.5	30.8	29.5	16.5	23.2	50.1	1.15:1	6.95	LFA
YU7EF EF7013M-6	3.83	16.11	1.33	1.25	22.01	4.7	32.9	26.4	18.7	21.8	49.4	1.06:1	6.84	Dipole
InnoV 14 LFA	4.03	16.46	1.38	1.32	22.31	4.5	30.0	30.8	16.0	22.3	52.6	1.14:1	7.42	LFA
*InnoV 14 LFA	4.03	16.46	1.36	1.29	22.26	4.5	30.0	30.6	16.0	22.3	52.6	1.14:1	7.40	LFA
+DG7YBN 14	4.06	16.38	1.38	1.31	22.30	3.5	30.7	26.9	15.9	20.6	49.4	1.51:1	7.42	Bent Dipole
YU7EF EF7014M-6	4.24	16.50	1.39	1.32	22.39	4.5	33.3	23.8	17.9	20.5	49.5	1.11:1	7.17	Dipole
RA3AQ AQ70-14f	4.30	16.63	1.41	1.34	22.54	6.9	33.6	26.1	15.9	20.0	52.8	1.09:1	7.28	Folded Dipole
KF2YN Boxkite 16	4.37	18.02	1.74	1.61	23.85	10.2	41.2	27.4	17.0	25.0	49.1	1.06:1	7.71	Bent Dipole
*KF2YN Polly 15 CR	4.44	16.76	1.40	1.40	22.66	3.5	30.2	26.7	16.3	20.3	47.4	1.21:1	7.86	Multi-pol loop
I0JXX 16JXX70	4.46	16.53	1.41	1.33	22.44	4.9	33.6	23.4	16.6	21.3	201.8	1.05:1	7.18	T Match
InnoV 15 LFA	4.59	16.68	1.42	1.34	22.57	4.8	30.0	31.1	15.8	22.1	51.1	1.08:1	7.81	LFA
YU7EF EF7015M-5	4.69	16.80	1.43	1.36	22.66	5.4	33.6	24.3	19.0	21.1	49.5	1.03:1	7.40	Dipole
WiMo 15 (YU7EF)	4.89	17.06	1.46	1.30	22.87	4.3	30.0	22.3	17.2	20.8	174.5	3.05:1	8.10	Folded Dipole

Hamsat - Amateur Radio Satellites Explained

TYPE OF ANTENNA	L λ	GAIN (dBd)	E (M)	H (M)	Ga (dBd)	Tlos (K)	Ta (K)	F/R (dB)	1st SL (dB)	2nd SL (dB)	Z (ohms)	VSWR Bandwidth	G/T (dB)	Feed System
InnoV 16 LFA	5.19	17.27	1.50	1.43	23.15	4.8	29.7	29.3	18.1	23.3	48.4	1.19:1	8.43	LFA
InnoV 17 LFA	5.33	17.25	1.49	1.42	23.11	4.3	29.6	27.4	18.1	23.5	49.9	1.03:1	8.40	LFA
*InnoV 17 LFA	5.33	17.25	1.45	1.45	23.11	4.3	29.5	27.4	18.1	23.5	49.9	1.03:1	8.41	LFA
YU7EF EF7017M-5	5.56	17.41	1.52	1.44	23.22	5.6	33.4	25.0	18.5	22.3	50.7	1.06:1	7.98	Dipole
*KF2YN Polly 18 CR	5.69	17.79	1.95	1.95	23.96	3.7	31.4	28.3	14.8	20	48.1	1.18:1	9.00	Multi-Pol loop
RA3AQ AQ70-18f	5.74	17.76	1.59	1.53	23.64	6.1	32.6	27.2	16.1	20.5	51.7	1.08:1	8.51	Folded Dipole
InnoV 18 LFA	5.76	17.49	1.52	1.45	23.33	4.3	28.9	28.2	18.5	24.8	49.8	1.05:1	8.72	LFA
+Konni F20 DL6WU	5.94	17.52	1.58	1.51	23.39	1.5	31.7	23.2	14.3	16.0	220.4	1.07:1	8.37	Folded Dipole
+DG7YBN 19	5.94	17.80	1.60	1.54	23.68	3.6	29.5	29.6	15.4	20.2	49.3	1.28:1	8.98	Bent Dipole
YU7EF EF7018M-5	6.00	17.61	1.54	1.48	23.39	5.1	31.7	26.4	17.7	22.6	49.3	1.08:1	8.39	Dipole
K1FO 22	6.10	17.85	1.63	1.56	23.75	4.8	33.6	20.6	15.4	21.2	200.8	1.34:1	8.49	T Match
YU7EF EF7019B-5	6.43	17.91	1.60	1.54	23.72	5.7	32.7	25.7	17.8	23.4	50.5	1.12:1	0.59	Dipole
KF2YN Polly 20 CR	6.50	18.30	1.90	1.90	24.43	3.8	31.1	28.7	14.4	19.5	48.8	1.21:1	9.50	Multi-Pol loop
InnoV 19 LFA	6.54	18.18	1.68	1.63	24.10	5.0	29.2	34.3	15.0	20.1	47.6	1.09:1	9.45	LFA
Tonna 21 DX	6.61	17.91	1.67	1.61	23.80	6.7	41.3	20.5	14.5	18.2	58.4	2.32:1	7.63	Folded Dipole
KF2YN Boxkite 22	6.70	19.26	1.99	1.90	25.14	10.8	40.2	29.0	16.2	21.2	52.1	1.17:1	9.11	Bent Dipole
InnoV 20 LFA	6.99	18.28	1.73	1.67	24.31	5.0	28.1	35.2	15.1	20.7	50.5	1.09:1	9.82	LFA
RA3AQ AQ70-21f	7.14	18.54	1.73	1.67	24.41	6.9	31.0	27.9	16.8	21.3	52.3	1.07:1	9.51	Folded Dipole
+DG7YBN 23	7.55	18.65	1.76	1.70	24.51	3.8	27.8	32.3	15.7	21.8	48.4	1.56:1	10.08	Bent Dipole
YU7EF EF7021B-5	7.59	18.66	1.74	1.68	24.50	7.8	32.3	29.6	20.3	21.7	50.2	1.33:1	9.41	Dipole
I0JXX 25JXX70	7.91	18.53	1.74	1.68	24.41	5.6	32.7	23.9	21.3	24.8	198.6	1.12:1	9.26	T Match
KF2YN Polly 24 CR	8.15	19.15	1.90	1.90	25.11	3.7	31.0	29.5	13.1	19.1	51.8	1.14:1	10.2	Multi-Pol loop
InnoV 23 LFA	8.40	19.16	1.87	1.82	25.08	5.6	28.2	31.8	15.9	21.7	45.2	1.23:1	10.58	LFA
*InnoV 23 LFA	8.40	19.16	1.89	1.85	25.06	5.6	28.2	31.8	15.9	21.7	45.2	1.23:1	10.56	LFA
DJ9BV BVO70-8.5wl	8.43	19.14	1.95	1.90	25.13	5.0	37.1	24.3	14.9	17.8	203.9	1.45:1	9.44	Folded Dipole
DJ9BV OPT70-8.5wl	8.44	19.04	1.80	1.75	24.87	4.9	33.9	23.2	15.8	20.1	186.1	1.56:1	9.57	Folded Dipole
YU7EF EF7023B-5	8.47	18.98	1.79	1.73	24.78	6.4	30.2	30.5	19.6	21.9	48.6	1.21:1	9.99	Dipole
RA3AQ AQ70-24f	8.52	19.22	1.87	1.82	24.82	6.1	29.0	30.0	15.1	21.0	52.0	1.13:1	10.20	Folded Dipole
InnoV 24 LFA	8.89	19.32	1.92	1.87	25.25	5.9	28.0	33.4	15.3	22.1	46.4	1.19:1	10.78	LFA
*InnoV 24 LFA	8.89	19.32	1.89	1.84	25.22	5.9	27.9	33.4	15.3	22.1	46.4	1.19:1	10.76	LFA
YU7EF EF7024B-5	8.90	19.19	1.84	1.79	25.00	7.6	31.5	30.3	20.5	22.4	50.1	1.60:1	10.04	Dipole
YU7EF EF7027B-5	10.18	19.67	1.94	1.89	25.47	7.8	32.2	25.6	20.0	23.2	50.3	1.64:1	10.39	Dipole

Ground Station Antenna

TYPE OF ANTENNA	L (λ)	GAIN (dBd)	E (M)	H (M)	Ga (dBd)	Tlos (K)	Ta (K)	F/R (dB)	1st SL (dB)	2nd SL (dB)	Z (ohms)	VSWR Bandwidth	G/T (dB)	Feed System
WiMo 27 (YU7EF)	10.43	19.43	1.85	1.80	25.08	4.3	27.5	27.1	20.5	22.6	191.8	2.72:1	10.70	Folded Dipole
K1FO 33	10.61	20.03	2.08	2.04	25.95	6.1	31.8	22.0	15.5	22.2	200.5	1.19:1	10.93	T Match
RA3AQ AQ70-30f	11.39	20.33	2.08	2.04	26.15	5.8	28.3	30.6	16.8	22.1	51.6	1.10:1	11.61	Folded Dipole
InnoV 30 LFA	11.64	20.45	2.27	2.22	26.42	6.7	30.6	36.6	13.9	20.3	49.6	2.38:1	11.56	LFA
*YU7EF EF7032-5	12.49	19.97	1.94	1.90	25.59	6.6	29.9	27.7	23.1	24.2	51.3	3.44:1	10.84	Dipole
InnoV 33 LFA	13.03	20.85	2.32	2.29	26.81	5.3	28.8	36.6	13.3	19.2	50.1	1.17:1	12.22	LFA
DJ9BV OPT70-13wl	13.29	20.85	2.35	2.29	26.80	5.2	33.4	25.9	14.8	19.7	183.0	1.44:1	11.56	Folded Dipole
I0JXX 39JXX70	13.53	20.16	2.00	1.97	25.86	6.3	30.9	26.1	26.7	27.0	194.0	1.19:1	10.97	T Match
InnoV 34 LFA	13.54	20.89	2.46	2.43	26.87	7.5	31.6	37.0	12.7	17.9	49.4	3.05:1	11.87	LFA
InnoV 35 LFA	14.00	20.95	2.46	2.43	26.93	7.5	31.4	36.7	12.7	18.1	49.1	2.69:1	11.96	LFA
InnoV 38 LFA	15.41	21.45	2.30	2.35	27.23	6.9	28.9	36.9	12.7	17.0	49.3	1.87:1	12.62	LFA
*InnoV 38 LFA	15.41	21.45	2.59	2.56	27.30	7.2	30.4	36.9	12.7	17.0	49.3	1.87:1	12.63	LFA
InnoV 40 LFA	16.23	21.67	2.69	2.66	27.67	7.5	29.4	37.8	12.6	17.2	48.0	2.27:1	12.99	LFA

Legend:
1. L = Length in wavelengths
2. Gain = Gain in dBi of a single antenna
3. E = E plane (Horizontal) stacking in meters.
4. H = H plane (Vertical) stacking in meters.
5. Ga = Gain in dBi of a 4 bay array
6. Tlos = Internal resistance of the antenna in degrees Kelvin.
7. Ta = Total temperature of the antenna or array in degrees Kelvin. This includes all the side lobes, rear lobes and internal resistance of the antenna or array.
8. F/R = Front to rear in dB over the rear 180 degree of an antenna using either E or H plane.
9. Z ohms = The natural impedance of a single antenna in free space.
10. VSWR = VSWR bandwidth is based a single antenna over 432.000 - 435.000MHz with a reference at 432.100MHz. This parameter gives an indicator of the antenna "Q" and what to expect with stacking and wet weather.
11. G/T = Figure of merit used to determine the receiving capability of the antenna or array = (Ga + 2.15) - (10*log Ta). The more positive figure, the better. G/T is modeled in Tant.exe at 30 degrees elevation.

Notes:
1. The programs used to calculate E/H Stacking, G, Ta, Tlos and G/T are EZNEC 5+ by Roy Lewallen W7EL, 4NEC2 by Arie Voors and Tant.exe by Sinisa, YT1NT/VE3EA. This combination of software provides excellent accuracy. Segment Density is 25 segments per half wave.
2. Temperatures used: Tsky=20 degree; Tearth=350 degree
3. Dipole Z is measured at 432.1MHz
4. F/R, 1st and 2nd Side Lobes (SL) have been calculated in a single antenna

5. No stacking harness loss or H frame effect are included in the 4 bay gain figures.
6. All stacking dimensions EXCEPT those marked with a "*" and "#" are calculated from the DL6WU stacking formula:

   $D = W/(2* \sin(B/2))$

Where:

   D = stacking distance, vertical or horizontal

   W = wavelength, in the same units as D

   B – beam-width between -3dB points.

   Use vertical beam-width for vertical stacking (as above),

   Use horizontal beam-width for horizontal stacking.

7. Antennas marked with a "*" have stacking dimensions recommended by the manufacturer or designer.
8. Antennas marked with a "#" have stacking dimensions for XPOL antennas by VE7BQH.
9. Antennas marked with a "+" have some or all elements over 6mm. All others are 4MM to 6MM.
10. FD = Folded Dipole
11. Manufacturer/Designer Legend:
12. Convergence Correction: NEC2 and NEC 4 are incapable of handling complex feed systems accurately like Folded Dipoles, T Matches, LFAs etc. Convergence Correction using the KF2YN Excel program is required. See DUBUS 4/2010 "The Correction of Convergence Errors in Antenna Temperature Calculations by Brian Cake, KF2YN for details.
13. Manufacturer/Designer Legend:

AF9Y = AF9Y	K2GAL = K2GAL
BVO = Eagle/DJ9BV	K5GW = Texas Towers/K5GW
BQH = VE7BQH	KF2YN = KF2YN
CC = Cushcraft	M2 = M²
CC MOD = VE7BQH	MBI = F/G8MBI/F5VHX
CD = CUE DEE	OZ5HF = Vargarda
CD MOD = VE7BQH	RA3AQ = RA3AQ
CT1FFU = CT1FFU	RU1AA = RU1AA
DD0VF = DD0VF	SHARK = SHARK (Italian)
DJ9BV = DJ9BV	SM2CEW = SM2CEW/VE7BQH
DJ9BV OPT = DJ9BV	SV = Svenska Antennspecialisten AB
DK7ZB = DK7ZB	
EKM MOD = SM2EKM	W1JR = VE7BQH (Mininec error)
F9FT = F9FT	WB9UWA = WB9UWA
Flexa = FlexaYagi	YU7EF = YU7EF
G0KSC LFA = G0KSC	UR5CSZ = UR5CSZ
G4CQM = G4CQM	UA9TC = UA9TC
HG = HYGAIN	Vine = G0KSC Design
I0JXX = I0JXX	YU7XL = YU7XL
IK0BZY = IK0BZY	InnoVAntennas = G0KSC
WiMo = WiMo	Tonna = F9FT
K1FO = K1FO	

Using this Chart:

While gain is important, other factors like easy matching and wet weather performance should be considered in the decision making. Antennas with 50 Ohm feed systems and good VSWR bandwidth (Q) may be the best choice depending on your location. Low side lobe and F/R antennas with good (G/T) may provide further significant benefit if you have local man-made noise that is in the direction where these kinds of antennas provide additional suppression.

Lionel H. Edwards

VE7BQH

Issue 2, December 14, 2012

The Endless Dilemma of Polarisation

What is the polarisation?

The electric field and the magnetic field of a radio wave at a certain distance from a transmitting antenna (for the principle of reciprocity, the same consideration shall be applied to a receiving antenna) are orthogonal to each other and propagate in the direction of the electromagnetic wave.

In the case of the so called, "linear polarisation", the two vectors propagate without rotating and we are talking about "vertical polarisation" or "horizontal polarisation", depending on the direction taken by the electric field E.

In case of circular polarisation, the two vectors, E and H, revolve around the axis of propagation of the wave, depending on the direction of rotation. Then we are talking about Right Hand Circular Polarisation RHCP or Left Hand Circular Polarisation LHCP. As in the case of linear polarisation, both the right and left circular polarisation are orthogonal in some way, in the sense that they can coexist without influencing each other.

This "no influence" is applicable only to an ideal antenna. In practice, any real antenna is not able to discriminate between the orthogonal polarisation in a perfect way: for example, a receiving antenna for vertical polarisation is also a little bit sensitive to a signal at the same frequency in horizontal polarisation. This phenomenon is measured by a parameter called "cross polarisation".

Linear or circular polarisation?

The theme of which polarisation to choose for our ground station antenna has always been one of the most debated among enthusiasts.

There is probably no perfect station, but simply the best compromise for a given type of traffic and cost/room tradeoff.

Let's see some factors to be considered:

- Some satellites have antennas with linear polarisation
- Some satellites have/have had antennas with right circular polarisation
- Some satellites have/have had antennas and left circular polarisation
- in the course of the orbit and under various conditions of propagation the polarisation plane wave varies due to:

    Faraday rotation in the passage through the ionosphere (especially at 145 and 435MHz)

    Mutual alignment of the antenna and the satellite antenna to the ground (attitude)

So a highly diversified pattern which makes the search for the "perfect solution" very complicated. Given the variables involved, the only certainty we have is that we will never receive a circularly polarised signal transmitted by a satellite with linear antennas. All other combinations are possible!

There are basically six possible approaches to the problem:

Solution	Schematic diagram
Antenna with fixed linear polarisation	Horizontal — Transceiver
Switchable H/V polarisation	Horizontal / Vertical switch — Transceiver
Fixed (right or left) circular polarisation	+45° and −45° elements → Power splitter — Transceiver
Switchable left/right circular polarisation	+45° and −45° elements → Power splitter — Transceiver (with length and length + 0.5 lambda switchable lines)

Ground Station Antenna

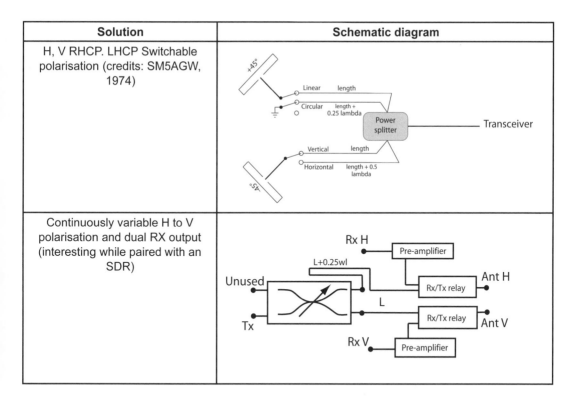

Note: some of the diagrams above are taken from the website: www.sm5bsz.com

Let's summarise the behaviour of the various solutions due to the variation of the polarisation of the incoming signal. In the table that follows, you can see the attenuation of the received signal compared to the best case, depending on the combination of antenna-polarisation versus signal polarisation.

Antenna configuration	Incoming signal polarisation				
	Horizontal	Vertical	RHCP	LHCP	Linear but with the wave plane shifted Φ degrees with respect the horizontal
Horizontal	0dB	> 20dB	3dB	3dB	$-10\log(\cos(\Phi))$
Vertical	> 20dB	0dB	3dB	3dB	$-10\log(\cos(90-\Phi))$
RHCP	3dB	3dB	0dB	> 20dB	3dB
LHCP	3dB	3dB	> 20dB	0dB	3dB

Please note also that:
- Circular polarisation is hardly ever found in the reality of amateur satellite traffic.
- As seen, it is relatively easy to change the polarisation, also in a continuous way, from the reception side on the ground in order to better adapt to that of the incoming signal. However, it becomes complex to do the same in the opposite direction, from our station to the satellite.

The right antenna for each frequency

Although all the antennas have the same purpose, there are many practical realisations. Their differentiation is due to mechanical limits at the extreme frequency (ie parts become either too large or too small, etc) there will be the different emphasis on the mix of benefits they offer (gain, G/T, bandwidth, polarisation, cost, etc).

Let's see in the following table (**Fig 8.9**) the most relevant to our traffic, broken down by popularity and typical application frequency bands:

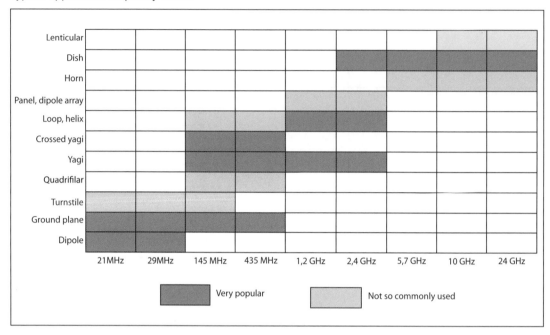

Fig 5.3: Most common kind of antenna versus amateur satellite band

The overall dimensions of the antenna

Once we have defined which antenna we wish to install, it is often necessary to determine whether the space they occupy is compatible with the room available in our garden or on our roof.

If the system is simple and compact, the problem does not pose particular difficulties. But when it becomes a bit more complex, verification prior to installation is recommended.

As a general rule, it would be desirable to maintain a minimum distance of at least one wavelength between each part of an antenna and other objects or obstacles, especially in the case of conductors. Consider as an example the general case of a group of antennas, the yagi type, which is equal in size, arranged at the vertices of a H structure, and free to rotate 360 degrees on the horizontal plane (azimuth movement) while 90 degrees toward the sky (zenith movement).

In general there are three risks of mechanical interference:

1. with any other antenna/mechanical structures nearby
2. with the floor/ground plane
3. with lines of wind-brace (if present)

Ground Station Antenna

Definitions	Illustrations
a = distance base/ground plane to the zenith rotation center b = distance from the azimuthal rotation centre and the external mast of the H structure c = distance from zenith rotation axis and antenna boom d = antenna length in front of the point of attachment to the mast e = antenna length behind the point of attachment to the mast f = half width of the antennas g = distance pivot point/ bolts (for wind-brace) attachment φ = azimuthal rotation angle λ = elevation angle γ = wind-braces angle with respect the support tower x = distance along the horizontal axis (right, left) y = distance along the orthogonal axis (front/rear) z = altitude along the vertical axis (ground to sky) Q = front-lower external antenna spot R = front-external top antenna spot T = rear-internal lower antenna spot Let's assume elevation 0 degrees to the horizon and 90 degrees pointing to the zenith. The azimuthal angle grows in a clockwise direction, similarly to what is indicated on the rotor.	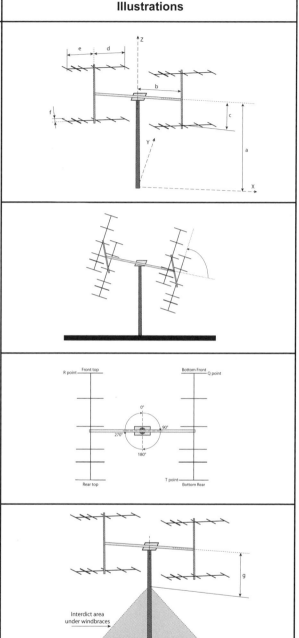

Let's begin with the simplest case, the first one.

We take as a common experience that the dimension "d" is greater than "e". In other words, the fixing point of the antenna is always shifted towards the tail in order to balance cable weight and the boom structure itself. In this condition, the antenna beaming horizon may interfere with its foremost element with other structures during the azimuthal rotation. To predict this event it is necessary to estimate the maximum size of the structure with respect to the vertical center of rotation, "S".

In this situation, the said distance is the projection on the base plane of the point R and therefore calculable by its Cartesian coordinates as:

$$S = \sqrt{x^2 + y^2}$$

From the definition of the characteristic points, we can write the equation of the projection of x and y according to the pointing and geometry considered:

$$x(\varphi) = b * \cos(\varphi) + d * \sin(\varphi) + f * \cos(\varphi)$$
$$y(\varphi) = b * \sin(\varphi) + d * \cos(\varphi) - f * \sin(\varphi)$$

As what is to be expected and demonstrated by a series of tedious mathematical manipulations, S does not depend on φ (pointing), but only on b, d, f according to:

$$S = \sqrt{(b+f)^2 + d^2}$$

This is the minimum required free space all around the vertical (tower, pole, etc) support in order to avoid any interference between antennas on the horizon and other obstacles.

You may ask what happens to the quota S if we raise the antennas to the sky. Let's see now this case, certainly very interesting for our application.

As before, we calculate the position on the x and y axes. This time not only the R point, but also on the lower homologous antenna, with "Q", as a function of the elevation angle λ.

For the sake of calculation simplification, while unrestrictive, we hypothesise that the pointing is fixed, with φ = 0 degrees.

For the point R:

$$x(\lambda) = b + f \rightarrow constant \; \forall \lambda$$
$$y(\lambda) = -c * \sin(\lambda) + d * \cos(\lambda)$$

while for the Q point:

$$x(\lambda) = b + f \rightarrow constant \; \forall \lambda$$
$$y(\lambda) = c * \sin(\lambda) + d * \cos(\lambda)$$

Now, given that we limit the field of the variable λ between 0 degrees and 90 degrees (horizon to zenith), it appears that:

$$c * \sin(\lambda) + d * \cos(\lambda) \geq -c * \sin(\lambda) + d * \cos(\lambda) \; \forall \lambda \in 0° \rightarrow 90°$$

replacing we have:

$$y[Q] \geq y[R] \; \forall \lambda \in 0° \rightarrow 90°$$

that for example gives:

$$S = \sqrt{(b+f)^2 + (c * \sin(\lambda) + d * \cos(\lambda))^2}$$

0 degrees elevation: (see above case) $\quad S = \sqrt{(b+f)^2 + d^2}$

90 degrees elevation: $\quad S = \sqrt{(b+f)^2 + c^2}$

with a simple spreadsheet you can see the trend of S as a function of λ.

For example, placing:
- b = 1.5
- c = 1m
- d = 3m
- f = 0.35m

you get the following pattern:

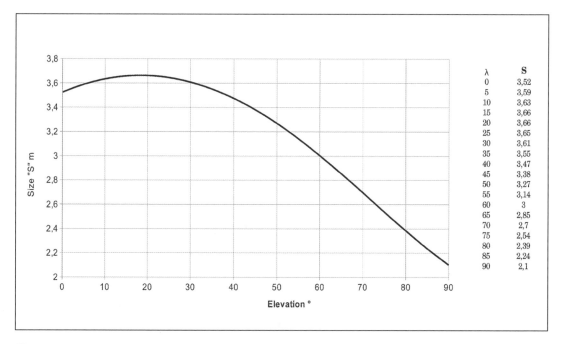

Fig 5.4: most common kind of antenna versus amateur satellite band

where it is visible the hump-shaped trend, with a maximum at around 15 degrees - 20 degrees and a value of S in the neighbourhood of 3.65m.

If the graphics solution does not meet the need of computational accuracy required, we have to proceed analytically, finding the maximum of the function before identified. Let's see how.

To find the local maximum of a function, the best method is to calculate its derivative function and look for zero value (zero slope equal to maximum or minimum).

Then, let's calculate the derivative function:

$$\frac{dS}{d\lambda} = \sqrt{(b+f)^2 + (c*\sin(\lambda) + d*\cos(\lambda))^2}$$

applying the usual rules of derivation, we obtain:

$$\frac{dS}{d\lambda} = c*\cos(\lambda) - d*\sin(\lambda)$$

being equal to zero we obtain the following:

$$\lambda_0 = \arctan\left(\frac{c}{d}\right)$$

applying this to the previous example, we obtain a value of 18.43 degrees of elevation with an associated value of S equal to 3.664m, really not far from the figures graphically calculated!

Frankly speaking, I always hope that there is enough clear room around the antennas in order not to have to make such sophisticated calculations!

Well, after checking the necessary room "around" the antennas, it is necessary to evaluate whether they are mounted at a sufficient height from the base so as to avoid interference with the ground.

As before, we calculate the position on the x and y axes of the T point (lower back of the antenna) as a function of the elevation angle λ. Even here, for the sake of calculation simplification, while unrestrictive, we assume that the pointing is fixed, with φ = 0 degrees. Since we are interested in a measurement along the vertical axis, the calculations will be developed along the z axis.

$$z(\lambda) = -c * \cos(\lambda) - e * \sin(\lambda)$$

as before, we can play with a spreadsheet to see the trend, assuming as example c = 1m and e = 2.5m, where the negative sign of S indicates that the footprint is downward below the elevation center of rotation. The function has a maximum of about 2.7m at 70 degrees elevation. As before, we can find the "precise solution", analytically:

$$\frac{dS}{d\lambda} = c * \cos(\lambda) - e * \sin(\lambda)$$

as being equal to zero, we obtain that the elevation of greatest downwards elongation of the antennas group becomes:

$$\lambda_v = \arctan\left(\frac{e}{c}\right)$$

which gives as a result a maximum z of -2.69m at 68.2 degrees of elevation. On the following, we will call this value Z_{max}

What has been seen now is the calculation for evaluating the minimum overall dimensions in all directions and pointing of our group of antennas.

In all the discussion conducted so far, it has been assumed that the support structure was self-supporting with negligible dimensions and shape compared with the space occupied by the antenna.

However, there is a quite common situation in which the above statement is not completely verified: if the vertical support requires wind-brace, the available space decreases and problems may arise.

Let's see now how to evaluate the possible interference between wind-brace and mechanical structure. Fortunately, many of the calculations already made come to help us.

From the initial graphics, we see that it is necessary to evaluate at the elevation of maximum vertical size (which we already know) if there is interference between the inner element of the antennas and the wind-brace.

If the value of Z of the bracing at a distance equal to (b-f) from the support, which we will call hereafter zT, is lower than Z_{max}, there will not be any problem.

Now let's see the equation:

$$zT(y,g) = -g - (b-f) * \cot(\gamma)$$

Now, for instance, let's assume to add one or more bracings, with a vertex angle of 30 degrees and anchored 1m below the centre of rotation of the group.

Ground Station Antenna

Let's see the result:

$$zT(30°,1) = -1-(1{,}5-0{,}35)*\cot(30°) = -2{,}99\,m$$

Z_{max} of the system was -2.69m and then between bracing and antenna elements will be there at least 30cm of room in the worst case. All right then!

Let's see in the table below the summary of the calculations:

Case	Illustration	Formulae
Horizontal footprint of the group to zero elevation		$S = \sqrt{(b+f)^2 + d^2}$
Horizontal footprint with elevation ranging between 0 degrees and 90 degrees		Elevation of maximum horizontal footprint: $\lambda_0 = \arctan\left(\dfrac{c}{d}\right)$ Maximum overall dimension $S = \sqrt{(b+f)^2 + (c*\sin(\lambda_0) + d*\cos(\lambda_0))^2}$
Vertical encumbrance (below the center of rotation) with elevation ranging between 0 degrees and 90 degrees		Elevation of maximum vertical dimension: $\lambda_v = \arctan\left(\dfrac{e}{c}\right)$ Maximum overall dimension $Z_{max} = -c*\cos(\lambda_v) - e*\sin(\lambda_v)$
Interference with the wind braces		No hassle if: $-g-(b-f)*\cot(\gamma) < Z_{max}$

Gallery of amateur antennas

Now, finally, some pictures of amateur solutions, from the most simple, inexpensive and portable to more complex and expensive.

Call sign	Set-up	Picture
GB4FUN (2012)	6 + 6 elements X-Quad @145MHz 9+9 elements X-Quad @ 435MHz 60cm dish @ 2400MHz	
IW4BLG (2009)	7+7 elements @ 145MHz, RHCP 144MHz mast mount preamplifier 18+18 element @ 43MHz, RHCP 435MHz mast mout preamplifier 4x10 elix for 1269MHz 144-1268 mast mout upconverter 1x0.8m net dish 2400MHz Patch feed G0RUH 2400-144MHz, UEK-2000 down converter	
I1TEX (2003)	17 elements @ 145MHz 21 elements @ 435MHz 55 elements @ 1269MHz 80x80cm dish @ 2400MHz 60cm dish @ 5700MHz 60 cm dish @ 10450MHz 40 cm dish @ 24050MHz	

Ground Station Antenna

Call sign	Set-up	Picture
Politecnico di Torino (2009)	2 x 9+9elements @ 145MHz 4x18+18 elements @ 435MHz 3m dish @ 2400MHz	
IW4BLG/p (2005)	HB9CV dual band	
I3ZJQ (2010)	9+9 elements @ 145MHz 2 x 9 elements @ 435MHz 60cm dish	
IW4BLG/3 (2003)	20 elements, log periodic 105-1300MHz	

Call sign	Set-up	Picture
IV3CYF (2010)	2 X 20 elements @ 435MHz, Home made 9 elements @ 145MHz, Tonnà 120 Cm prime focus dish @ 2.4GHz Mast mount 435MHz preamplifier Mast mount 145MHz preamplifier (home made)	
IK1ODO (2010)	2 x 9 el. @ 145MHz 2 x 23 el. @ 435MHz 60cm dish @ 2400MHz	
IW4BLG/3 (2004)	5+5 elements @ 145MHz, RHCP 19+19 elements @ 435MHz, RHCP 4x10 elix @ 1296MHz	

Reference:
1) EZNEC 5+ by Roy Lewallen W7EL,
2) 4NEC2 by Arie Voors and Tant.exe by Sinisa, YT1NT/VE3EA

The Doppler Effect

The doppler effect is an apparent change in frequency or wavelength perceived by an observer who is in relative motion with respect to the source. It owes its name to Christian Andreas Doppler, who first studied this effect in 1845. Doppler's experiment became famous: he stood next to the railroad tracks and listened to the sound emitted by a compartment full of musicians, hired for the occasion, while the train approached and then moved away. The test confirmed that the pitch was higher when the source of the sound was getting closer and lower when it was going away, exactly what he had predicted.

It is important to note that the radiation frequency does not change in the reference system rigid with the source.

To understand the phenomenon with a concrete example, consider the following analogy: when we are standing on a beach, we get the waves from the sea with a certain rhythmic flow. Getting into the water and moving towards the open sea (considered the source), we will face the waves, so we will meet them more frequently (frequency increases). But when we get back to the shore, walking in the same direction of the waves, the frequency will decrease.

Doppler estimated this effect mathematically as follows:

$$f = f_o \cdot \left(\frac{v + v_m - v_o}{v + v_m - v_s} \right)$$

Where:

- v = velocity of the waves in the medium
- v_m = velocity of the medium
- v_s = velocity of the source
- v_o = velocity of the observer

The speed is considered positive if it is in the same direction along which the wave propagates or negative if in the opposite direction.

The formulation given by Doppler (said standard Doppler), however, is applicable only to the waves that require a means to propagate and move with a non-relativistic speed, ie much slower than the light.

In his special theory of relativity, Einstein introduced a new formula that describes the Doppler effect of electromagnetic radiation such as radio waves, which do not need a means to propagate and do so at speed approaching or equal to those of light.

Highlighting the frequency or wavelength, the formula can be written as follows:

Frequency	Wavelength
$$v_o = \sqrt{\frac{1-\frac{v}{c}}{1+\frac{v}{c}}} * v_s$$	$$\lambda_o = \sqrt{\frac{1+\frac{v}{c}}{1-\frac{v}{c}}} * \lambda_s$$
v_o = observed frequency	λ_s = wavelength emitted
v_s = emitted frequency	λ_o = wavelength observed
v = relative velocity of the source	v = relative velocity of the source
c = speed of light (299,792.458km/h)	c = speed of light (299,792.458km/h)

Effects of the Doppler on satellite service

Unlike point-to-point links between fixed earth stations, in which the received signal remains substantially stable in frequency, in satellite links the Doppler Effect leads to a continuous change of the received frequency (the only exception would be with the use of geostationary satellites).

Let's see in the following table the maximum value of the frequency shift for a generic satellite in LEO orbit at an altitude of approximately 800km as a function of the various frequency bands used:

Band	15m	10m	2m	70cm	23cm	13cm	3cm
Freq. (MHz)	21.280	29.400	145.900	435.070	1269.000	2401.000	10450.000
Max Doppler	+/- 477Hz	+/- 659Hz	+/- 3.27kHz	+/- 9.76kHz	+/- 28.5kHz	+/- 53.8kHz	+/- 230kHz

As has already seen from the mathematics, the Doppler Effect increases with increasing frequency. In the case of narrow band traffic, like telegraphy or SSB, this effect will be perceptible in almost every band. In the case of FM voice or data traffic, up to 145MHz, it is possible to work without adjusting the frequency of the receiver and the transmitter.

Especially when using an FM repeater or when communicating near the edge of a linear transponder, it should be noted that the Doppler Effect influences both up-link and the down-link.

The shift in the down-link is instinctively compensated by the operator by adjusting the frequency of the receiver to keep it well tuned.

More "subtle" is the effect on the up-link. The frequency shift received from the satellite could slide our signal outside of the pass band of the transponder or generate serious distortion in the case of FM.

However, the Doppler Effect itself is not the worst enemy of amateur communications satellites, other than its constantly changing with non-linear law, especially in orbits high above our horizon.

See for example the diagram in **Fig 6.1**.

The Doppler Effect

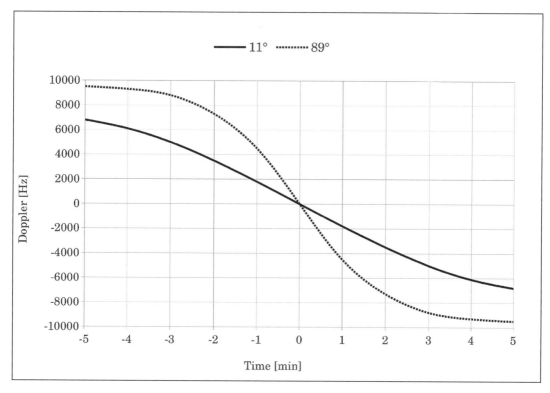

Fig 6.1: Trend of the Doppler Effect at different max elevation of the pass

It describes the Doppler Effect variation at two passages at different maximum heights above the horizon, referring to the instant of maximum elevation.

Its compensation requires in these cases "good hand and trained ear" or alternatively, a good computerised system of calculation and automatic correction of the transceiver.

Today, most radios used by radio amateurs are able to offer this feature once interfaced with a personal computer running one of the various tracking programs. This will be discussed later on.

Inverting and non-inverting transponder

Let's see now how we can mitigate the Doppler Effect to some extent on our satellite communications.

For example, there is a satellite in LEO orbit which operates J or VU mode with up-link on 2m band and downlink on 70cm.

Let's assume that the satellite has a linear transponder capable of converting an input of 145.900MHz to a signal retransmitted to the ground at 435.500MHz.

The block diagram could be the following (**Fig 6.2**).

There are two ways to convert the input signal.

Summing up the input signal with the local oscillator (which then will be lower than the output frequency), or subtracting the frequency of the local oscillator (hence higher than the output frequency) the input signal.

Hamsat - Amateur Radio Satellites Explained

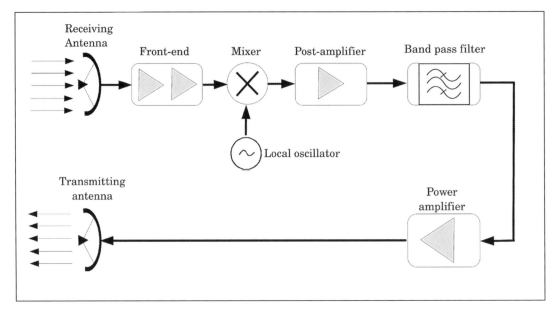

Fig 6.2

The table below shows the results:

	Case A	Case B
Local oscillator frequency	289.60MHz	581.400MHz

Let's see how the two transponders behave against the frequency of the input signal:

Input Frequency [MHz]	Output Frequency [MHz] (Case A: Fdwn = Fol + Fup)	Output Frequency [MHz] (Case B: Fdwn = Fol-Fup)
145.880	435.480	435.520
145.890	435.490	435.510
145.900	435.500	435.500
145.910	435.510	435.490
145.920	435.520	435.480

From the table, it can be observed that in case A the increase of the input frequency leads to the increase of output frequency while in case B the opposite happens as visible also in **Fig 6.3**.

These different behaviours are called "non-inverting" and "inverting". It is worth mentioning as in the case of single-sideband (SSB) signals, that **Case A** (non-inverting), save the mode between (USB->USB or LSB->LSB), while in **Case B** (inverting) modes are inverted (USB->LSB or LSB->USB).

The Doppler Effect

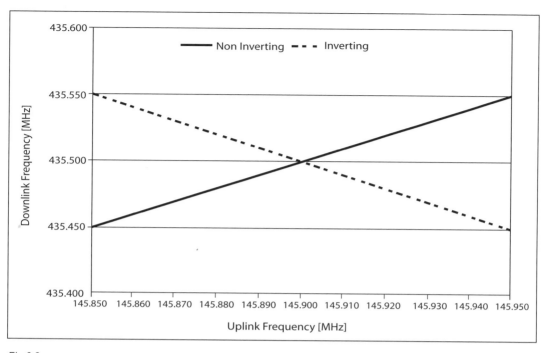

Fig 6.3

Instinctively, one would think that A is more advantageous: the local oscillator in the satellite works at a lower frequency, it saves the SSB mode and it's more instinctive to change the tuning in the ground station (if you increase the uplink frequency, you have to go into the same direction with the receiver).

But how are the two systems with regard to the Doppler Effect?

Here is a satellite shortly after the acquisition and with a relative speed towards us of about 7km per second. Assume also that at that time it is in connection with another station that instead has the satellite almost set, then going away at the same speed.

The table below shows the results:

	Case A, non-inverting	Case B, inverting
Up-link frequency [MHz]	145.900000	
Frequency as received by the satellite [MHz]	145.903407	
Frequency of the local oscillator of the transponder [MHz]	289.600	581.400
Output frequency from the satellite [MHz]	435.503407	435.496593
Downlink frequency as received on our station [MHz]	435.513576	435.506762
Downlink frequency as received on correspondent station [MHz]	435.493238	435.486425
Doppler for us [kHz]	13.576	6.762

	Case A, non-inverting	Case B, inverting
Doppler for the corresponding [kHz]	-6.762	-13.576

From the table you can see that, in the case of inverting transponder, the playback of our emission is affected by a Doppler shift which is halved compared to the case of non-inverting transponder.

The situation is changed by analysing the point of view of the hypothetical correspondent who was in the opposite situation with regard to the relative motion of the satellite.

The general trend is a preference of the former situation, which ensures a reduced Doppler Effect on the reception of its own signal.

For that reason there are many amateur satellites with an on-board transponder of the inverting type.

Frequency and "Mode" of Amateur Satellites

The National Frequency Band Plan

All amateur satellites and the ground stations are subject to the management and coordination of the frequencies in force in each country which implements the Radio Regulations of the ITU in its own legal system.

There are three types of assignments (status) of frequency band:

Exclusive: the frequency band is allocated exclusively to the indicated service and no other service can use it.

Primary: the frequency band is shared with other services called secondary.

Secondary: the frequency band is shared with other services such as primary.

A secondary service station:
- must not cause harmful interference to stations of a primary service to which the frequencies are already assigned or can be assigned;
- cannot claim protection from harmful interference caused by stations of a primary service to which frequencies are already assigned or can be assigned;
- is entitled to protection from harmful interference caused by stations in the same or ancillary services to which frequencies can then be assigned;

It is almost impossible to show in this book a detailed and up to date description of the situation in each country. However, to help the reader, let's check some definitions related to our service and used within the rules all around the world.

Allocation (of a frequency or channel) – Permission granted by the competent body for the use of a certain frequency or a radio channel in a radio station subject to specific conditions.

Allocation (of a frequency band) – Entry in the national plan for the allocation of frequencies in a frequency band determined for the purpose of use by one or more earth or space radio services or from the radio astronomy service.

Detrimental disturbance – Disturbance which affects the functioning of a radio navigation service or other safety service or seriously degrades, repeatedly interrupts or prevents the operation of a radio communications service used in accordance with this plan.

Radio waves – Electromagnetic waves whose frequency is conventionally less than 3000GHz, propagating in space without artificial guide.

Radio communication – Telecommunication achieved by means of radio waves.

Earth Radio communication – Any radio communication other than space radio communications and radio astronomy.

Space Radio communication – Any radio communication by means of one or more space stations or by means of one or more satellites or other reflective object in space.

Radio Astronomy – Astronomy based on the reception of radio waves of cosmic origin.

Amateur Service – Radio communications service having as its purpose individual instruction, intercommunication and technical studies carried out by amateurs who are duly authorised persons interested in the technique of radio electronics without personal pecuniary interest.

Amateur Satellite Service – Radio service using space stations or satellites for the same purposes as the amateur service.

Radio Astronomy Service – Service involving radio astronomy use.

Radio Communications Service – Service involving the transmission, emission or reception of radio waves for specific telecommunication purposes.

Fixed Service – Radio communications service between fixed points.

Fixed-Satellite Service – Radio communications service between ground stations located in certain positions performed by means of one or more satellites. This service may also include feeder links for other radio services.

Mobile Service – Service of radio communications between mobile stations and land stations or between mobile stations.

Mobile-Satellite Service – Service of radio communications between mobile earth stations and one or more space stations or between space stations used by this service or between mobile earth stations by means of one or more space stations. This service may also include feeder links necessary for its operation.

Land Mobile Satellite Service – Mobile satellite service in which mobile earth stations are located on land.

Radio Station – One or more transmitters or receivers or the combination of transmitters and receivers including the accessory equipment, all of which are necessary in a given location to ensure radio communications service or the radio astronomy service. Each station is classified on the basis of the service it participates in, in a permanent or temporary way.

Earth Station – Radio station located on the earth which provides radio communication service. Unless otherwise stated, this term has the same meaning as ground station.

Ground Station – Station located on the earth's surface or at least on the main part of the atmosphere and intended to communicate with:

- One or more space stations.
- With one or more stations of the same kind by means of one or more satellites or other reflective objects in space.

Space Station – Station located in an object that is destined to go or has gone beyond the main part of the earth's atmosphere.

Mobile Station – Station of the mobile service intended to be used when it is moving or parked in undetermined points.

Ground Mobile Station – Mobile station on the earth for satellite service when moving or parked in undetermined points.

Base Station – Earth station of land mobile service.

Radio Astronomy Station – Station of radio astronomy service.

Telecommunications – Any transmission, emission or reception of signs, signals, writing, images, sounds or intelligence of any nature by wire, or carried out by radio waves, optical or other electromagnetic system.

Industrial, Scientific and Medical Uses (ISM) – Implementation and installation of equipment designed to produce and use, in a small space, radio energy for industrial, scientific, medical, domestic or similar, with the exclusion of any use of telecommunications.

Frequency and "Mode" of Amateur Satellites

The IARU Band Plan

The IARU (International Amateur Radio Union) describes how to use each frequency band granted to the amateur service. The latest revisions took place at the IARU Region 1 Conference.

The band plan is the "law" to which radio amateurs adhere in order to ensure that everybody the maximum enjoyment of the frequencies assigned to all the different services that we use as amateur radio enthusiasts.

Band Plan for Space Communications:

Frequency MHz	Max bandwidth Hz	Note
29.300 – 29.510	6000	Only space to ground communications
145.794 – 145.806	12000	FM or digital voice. Note 1
145.806 – 146.000	12000	All modes, note 1
145.194 - 145.206	12000	Recommended frequencies for voice communication FM to /from space manned missions i.e. ISS
435.000 – 438.000	20000	Max 20kHz of bandwidth
1260 - 1270		Only ground to space communication
2400 – 2450		
5650 - 5668		Solo up-link (terra-spazio)
5668 – 5670		Only up-link shared with narrow band ground traffic
5790 – 5850		Only downlink
10450 - 10500		Shared with ground services
24048 - 24050		Shared with narrowband ground services
47088 - 47090		Shared with narrowband ground services
75500 - 76000		Shared with ground services
77500 - 77501		Shared with narrowband ground services
134000 - 134001		Shared with narrowband ground services
248000 - 248001		Shared with narrowband ground services

Note 1: The 2m band is one of the most intensely used for amateur satellite operations. It has an exclusive and globally coordinated sub-band at 145.800-146MHz

Satellites Modes

The satellites for telecommunications typically have one or more transponders on board, ie devices capable of receiving the signal from the ground of a given frequency (or band) and retransmitting that to another frequency or band.

The pairing of bands or frequencies used for up-link and down-link, is called the "mode" of the satellite.

Over the years, the name has undergone a significant change. The first version, actually still in use in the common lexicon, pointed out the mode with a single capital letter (ie A, B, J, etc) according to the historical evolution of the transponders.

The new nomenclature, instead, makes use of a pair of upper case letters, which respectively indicate the band of the up-link and down-link according to the following encoding. It is therefore independent from the historical sequence of appearance.

The various frequency bands are briefly identified as follows:

Band	Frequency	Code
HF	21 – 30MHz	H

Hamsat - Amateur Radio Satellites Explained

Band	Frequency	Code
VHF	144-148MHz	V
UHF	438 – 438MHz	U
	1260 – 1270MHz	L
	2400 – 2450MHz	S
SHF	5800MHz	C
	10450MHz	X
	24050MHz	K

In the following summary table, you can find those used so far with both the new and the old nomenclature:

Old	New	Up-link [band, MHz]	Down-link [band, MHz]	Note
A	V/H	2m, 145.800 – 146.00	10m, 29.00 – 29.500	
B	U/V	70cm, 435.00 – 437.50	2m, 145.00 – 146.00	
T	H/V	15m, 21.50 – 21.00	2m, 145.00 – 146.00	Only Soviet satellites
K	H/H	15m, 21.150 – 21.300	10m, 29.300 – 29.500	Only Soviet satellites
L	L/U	24cm, 1267.000 – 1270.000	70cm, 435.000 – 437.150	
S	U/S	70cm, 435.000 – 437.150	13cm, – 2400.000 – 2401.500	
JA	V/U	2m, 145.800 – 146.000	70cm, 435.000 – 437.150	Mode J, analog
JD	V/U	2m, 145.800 – 146.000	70cm, 435.000 – 437.150	Mode J, digital
	L-U	24cm, 1267.000 – 1270.000	70cm, 435.000 – 437.150	
	H/S	10m, 29.300 – 29.500	13cm, – 2400.000 – 2401.500	
	L/S	24cm, 1267.000 – 1270.000	13cm, – 2400.000 – 2401.500	
	L/X	13cm, 2400,000 – 2401,500	3cm, 10450.000	
	C/X	6cm, 5800.000	3cm, 10450.000	
	L/K	24cm, 1267,000 – 1270,000	1.5cm, 24050.000	OSCAR 40

Important note: the frequencies mentioned in the table are those assigned into the band plan and do not necessarily express those actually used by each satellite (generally more restricted).

The mode evolution has been driven substantially by two factors:
- to broaden user base
- to offer new opportunities for study and experimentation

The Keplerian Elements

In order to indicate and predict the position of a satellite in space, we rely on a set of information called orbital elements. These are numbers that uniquely define the orbit of each satellite and are generally advertised in various ways through mailing lists, newsletters, magazines and websites. Before moving into the study of these parameters, it is worth mentioning briefly the three main laws that describe the orbit of a celestial body, with thanks to the great physicist and astronomer Johannes Kepler.

Kepler's First Law

"The orbit of a planet is an ellipse with the sun at one focus"

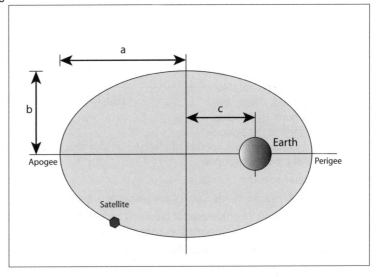

Fig: 8.1 Graphical representation of the Kepler's First Law

The great revolution of Kepler takes shape in his first statement sweeping away the celestial spheres and the perfect circular orbits by introducing elliptical orbits instead. The motions of the planets and satellites take place in a plane which is called the orbital plane.

A generic elliptical orbit of a satellite around the earth indicates some of the following features:

    a = semi-major axis
    b = semi-minor axis
    c = semi-focal distance
    e = eccentricity

As per the geometry basis, the laws that bind them to each other are as follows:

$$c = \sqrt{a^2 - b^2}$$

and

$$e = \frac{c}{a}$$

It is a common sense to consider the First Kepler Law to be connected to the conservation of momentum.

Kepler's Second Law (1609), The Law of Areas

The vector joining the centre of the sun with the centre of the planet sweeps equal areas in equal times. A graphical representation is shown in **Fig 8.2**.

In practical terms, Kepler's Second Law expresses the concept that if the segments AB and CD are covered in equal times, the areas of the "orbital slices" AB-Earth and CD-Earth are equal.

Let's see now some consequences of this law:

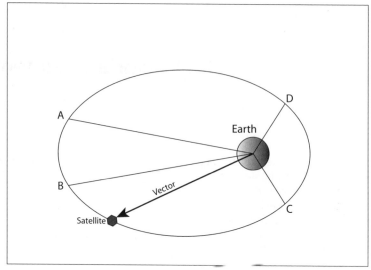

Fig: 8.2 Graphical representation of the Kepler's Second Law

- In the most common case of non-circular orbit, the orbital velocity of the satellite is not constant. Near the perigee, where the radius is shorter than at the apogee, the arc of ellipse is correspondingly longer. In turn this means that the orbital speed is highest at perigee and lowest at apogee.
- The areolar speed is constant.
- The orbital angular momentum of the planet is preserved (see below)
- The speed along a certain orbit is inversely proportional to the magnitude of the radius vector. This is a consequence of angular momentum conservation.

A central force is exerted in the satellite, directed along the line joining the satellite and our planet.

Kepler's Third Law (1619)

'The square of planet revolution period is proportional to the cube of the semi-major axes of their orbits'.
This law can be expressed in mathematical form as follows:

$$K = \frac{T^2}{d^3}$$

Where:
- d = semi-major axis
- T = revolution period
- K = constant (sometimes called Kepler) which depends on the celestial body around which occurs the motion of revolution.

In the case of a circular orbit, the formula becomes:

$$T^2 = r^3 \cdot K$$

Where r is the radius of the orbit.

The Keplerian Elements

Limits of Validity of Kepler's Laws

It should be noted that the application of Kepler's laws to our satellites is reliable, until the following assumptions are fulfilled:
- The satellite mass is negligible compared to that of the earth (almost always true)
- We can neglect the interactions with other celestial bodies (sun and moon which slightly perturb the orbits of our satellites)

The seven (or eight) Keplerian elements

Seven numbers are necessary to fully define the orbit of a satellite. This group of numbers is called "orbital elements" or more commonly "Keplerian elements" in memory of the great astronomer Johann Kepler [1571-1630].

These parameters define an ellipse oriented with respect to the earth and where the satellite is placed at a defined time. Unfortunately, the reality is a little more complex than what Kepler believed. There are, in fact, at least two elements that disturb this ideal situation: the inhomogeneity of the gravitational field and the friction with the upper atmosphere. The first one is usually taken into account by the program for the calculation of ephemeris, while the latter one may, in some cases, become the eighth issue of the family of Keplerian parameters.

Now let's see the list of the "fantastic seven (eight)":

1. Epoch
2. Orbital Inclination
3. Right Ascension of Ascending Node (RAAN)
4. Argument of Perigee
5. Eccentricity
6. Mean Motion
7. Mean Anomaly
8. Drag (optional)

1 Epoch

Also called "Epoch Time" or "T0"

This Keplerian parameter is similar to a "snapshot" taken at a given moment in the life of the satellite. This figure defines the moment when the group of parameters is reported.

2 Orbital Inclination

It is also called "Inclination" or "I0".

Each closed orbit lies in a plane known as the orbital plane which passes through the centre of our planet. This parameter indicates its inclination with respect to the orbital plane (ecliptic) of the earth.

By definition, it can vary from 0 degrees (when the satellite orbit is around the equator) to 180 degrees (specular situation). Orbits with an inclination close to 90 degrees are called polar as the satellite flies right over the two poles of the earth.

The intersection of the orbit plane with the equatorial one is called line of nodes which will be discussed later.

3 Right Ascension of Ascending Node

It is also called "RAAN" or "RA Node" or "O0" or even "Longitude of Ascending Node".

Together with the inclination, it describes and identifies the orbital plane.

Due to the rotation of the earth, it is not possible to use terrestrial coordinates to indicate the direction of the line of nodes. Instead, we have to employ the celestial coordinate system used by astronomers known as right ascension and declination which does not rotate with the Earth.

The right ascension of a satellite, or more generically of a celestial body, is the angular distance between the celestial meridian passing through the satellite in question and another celestial meridian of reference that will pass through the point of intersection (ascending node) between the equator celestial and the ecliptic. In practice, the reference meridian passes through the point taken by the sun on the day of the vernal equinox.

In other words, the "right ascension of ascending node" is the angle measured at the centre of our planet between the spring equinox and the ascending node.

We have just talked about the spring equinox which astronomically is also called the vernal point. Known as the First Point of Aries or gamma point (γ is similar to the symbol that marks the constellation of Aries), it is one of the two equinoctial points where the celestial equator intersects the ecliptic.

Here is an example: suppose you draw a line from the center of the Earth to the point where the satellite, going from south to north, crosses the equator. If this line points to the spring solstice, then RAAN is 0 degrees. By convention, RAAN is defined between 0 degrees and 360 degrees.

4 Argument of Perigee

It is also known as "ARGP" or "W0".

It expresses the orientation of the orbital ellipse by means of an angle.

To fully understand its meaning, we need a little insight into the elliptical orbits. The point on the orbit of minimum distance from the earth is called perigee, or periapside or perifocus.

Similarly, the opposite point, the maximum distance is called apogee or apoapside, or apifocus.

The imaginary line joining apogee and perigee is called the line of apsides or major axis of the ellipse and of course it passes through the center of our globe.

We also have to remember that the intersection of the equatorial plane with the orbital plane is called the line of nodes. The angle between these two lines is called the argument of perigee.

Obviously, any intersection of straight lines forms a pair of supplementary angles. So for complete definition, the argument of perigee is the angle measured in the centre of the earth between the ascending node and the perigee.

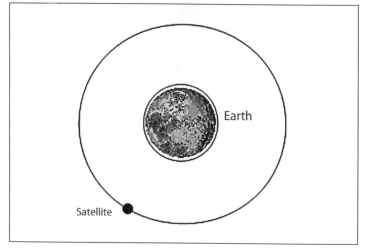

Fig 8.3: Circular orbit e=0

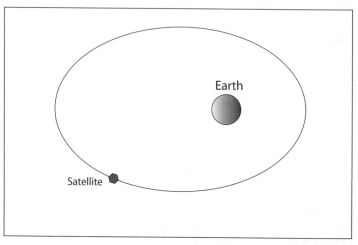

Fig 8.4: Elliptical orbit 0<e<1

Let's look at two examples:
- argp = 0 degrees the perigee is coincident with the ascending node. The satellite will pass closest to the Earth right on the equator line.
- argp = 180 degrees the apogee is coincident with the ascending node. The satellite will go as far away right on the equator of the Earth.

By definition, the argument of perigee is an angle variable between 0 degrees and 360 degrees.

5 Eccentricity

Also known as "ecce" or "E0" or "e", it is the parameter that indicates how the orbit "differs" from a perfect circle. The orbital eccentricity is strictly defined for all kinds of orbits as circular, elliptic, parabolic and hyperbolic:
- for circular orbits: e = 0   (see **Fig 8.3**)
- for elliptic orbits: 0 <e <1   (see **Fig 8.4**)
- for the parabolic paths: e = 1   (see **Fig 8.5**)
- for hyperbolic trajectories: e> 1   (see **Fig 8.6**)

For elliptical orbits, typically those of our amateur satellites, the eccentricity can be calculated from apoapsis and periapsis as follow:

$$e = \frac{d_{(a)} - d_{(p)}}{d_{(a)} + d_{(p)}}$$

where:
- $d_{(p)}$ is the radius of pericentre, ie the radius of the perigee, calculated in relation to the terrestrial centre
- $d_{(a)}$ is the radius of apocenter, ie the radius of the apogee, calculated in relation to the earth's centre

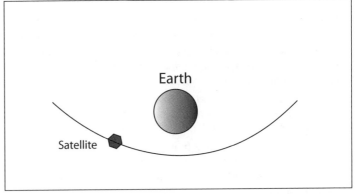

Fig 8.5: Parabolic orbit e=1

6 Mean Motion

Also known as "N0", it expresses the amplitude of the orbit by means of the average speed of the satellite.

We know how Kepler's Third Law expresses the relationship between orbital speed and distance from the Earth. In the case of circular orbits the proportion is constant and in the case of elliptical orbits, the speed varies (minimum at apogee and maximum at perigee) and then we use the average speed which, in astronomy, is called "Mean Motion". The Mean Motion is generally measured as

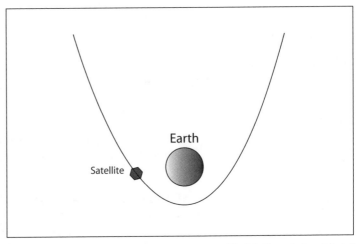

Fig 8.6: Hyperbolic orbit e>1

the number of revolutions per day, which means the time elapsed between passes through the perigee. Sometimes in its place the term "orbit period" is used, which is nothing but the reciprocal of the Mean Motion. For example, a satellite with a Mean Motion of 4 revolutions per day has an orbital period of 6 hours!

The typical values for amateur satellites may vary from 1 rev per day to about 16 revs per day.

7 Mean Anomaly

Also known as "M0" or "MA" or "Phase", it indicates the position of the satellite in its orbit expressed by means of an angle that is 0 degrees at perigee and 180 degrees at apogee.

In the ideal case of a perfectly circular orbit, which is characterised by a constant orbital speed sitting in the centre of the earth and beaming toward the angle Mean Anomaly, we will be facing the satellite. In the real case of elliptical orbits, where the speed of the satellite varies, this does not happen, for two reasons, the perigee will always be at MA = 0 degrees and the apogee always at MA = 180 degrees regardless of eccentricity.

It is a common practice to use the data of MA to schedule the satellite operation. In this case, the MA is expressed on a scale between 0 (equal to 0 degrees) and 256 (for the 180 degrees) and no longer in degrees. To highlight this fact, many tracking programs use the term "phase" to indicate the Mean Anomaly in this special way.

8 Drag

Also called "N1", it indicates the effect of the friction caused by the high atmosphere which tends to slow down the satellite and place it into a re-entry spiral orbit (destructive in the long term).

By definition it is the half of the first derivative of the term Mean Motion. In other words, it expresses the rate of change of MM due to the effect of aerodynamic drag. Its value, always very small, is measured in revolution per day. The typical values are around 10^{-4} for LEO satellites and 10^{-7} for those HEO.

Although there is no doubt that the interaction with the terrestrial atmosphere lowers the orbit of the satellite, this value can occasionally be published as negative. The reasons for this phenomenon, which at the first glance seems absurd, are:

- Errors due to measurement imprecision. The parameter is measured in an indirect manner from other orbital parameters and particularly when it has values close to zero the measurement errors can lead to slightly negative results.
- The satellite is subject not only to the earth's gravitational field and the friction with the atmosphere. The gravitational field of the sun and moon will also disturb its run. Especially when it is aligned, there may be a slight and temporary increase of the orbital altitude. In summary, if the parameters have to be used for short term forecast, it is worth considering also negative drag, while if the goal is an estimation of the orbital position over the medium to long term, it is more convenient to use only drag values equal to or greater than zero.

Other parameters

There are other optional parameters that can allow the tracking programs to provide more information for the users.

Epoch Rev

Also called "Revolution Number at Epoch", this figure indicates how many orbits the satellite has completed at the time, as "Epoch". Epoch Rev is used to display the number of the current orbit.

Attitude

Also known as "Bahn Coordinates", it expresses the orientation of the satellite in space and may actually indicate whether its antennas are pointed at our station or not.

There are various methods to describe and indicate the attitude but the most common one is the Bahn Coordinates, applied to the satellites equipped with stabilisation by means of rotation which maintains a constant orientation for inertial effect (such as the axis of a spinning top for example). The Bahn Coordinates are expressed by means of two angles, the so-called Bahn Latitude and Longitude representing a spherical coordinate system with the earth as the centre. The main axis of this coordinate system is the one that connects the centre of the earth with the perigee point of the satellite.

Format of the Keplerian Elements

The orbital elements for amateur satellites are commonly distributed in two typical forms:

NASA 2-Lines

This is the format used by NASA to distribute the Keplerian elements in the "NASA Prediction Bulletin".

For amateur satellites, they are composed of three lines of data: the first contains only the name of the satellite while the two following (hence the name "2-lines") contain all the orbital information.

A typical format NASA looks like this:

OSCAR 10
```
1 14129U    88230.56274695 0.00000042      10000-3 0 3478
2 14129     27.2218 308.9614 6028 281 329.3891    6.4794 2.05877164 10960
```

Each number has a fixed position into the string, the spaces are significant and the rightmost digit is a decimal-based control character of data integrity.

The tracking program usually also controls the sequence and performs a plausibility test on each of the values, ensuring high safety and reliability.

The block of data has the following structure:

```
AAAAAAAAAAA
1 NNNNNU NNNNNAAA NNNNN.NNNNNNNN +.NNNNNNNN +NNNNN-NNNNN-N +N N NNNNN
2 NNNNN NNN.NNNN NNN.NNNN NNNNNNN NNN.NNNN NNN.NNNN NN.NNNNNNNNNNNNN
```

Heading

> It is the name of the satellite expressed with 11 digits. There are also options to 12 digits and certain versions to 24 digits as reported by NORAD documentation.
>
> This line has no control letter.

The first line
It is built according to the following rules:

Columns	Description
01-02	Number of line
03-04	Number of satellite
10-12	International Designator (last two digits of the year of launch)
13-14	International Designator (Launch number of the year)
15-17	International Designator
19-20	Epoch Year (last two digits of the year)
21-32	Epoch (integer and fraction of a day)
34-43	First derivative of Mean Motion divided by 2, or ballistic coefficient (depending on the type of ephemeris)
45-52	Second derivative of Mean Motion divided by 6 (white spaces if not available)
54-61	Term friction "bstar" when using the general theory of perturbations GP4, or radiation pressure coefficient.
63-63	Ephemeris type
65-68	Element number
69-69	Checksum

The checksum is calculated as follows:

- start at zero
- for each digit along the line, adds its value
- for each "minus" sign, adding up the value one
- for each "plus" sign, adding up the value two (sometimes zero depending on the "compiler" of the data)
- for each letter, space or comma, sum zero

The least significant digit of the sum, in decimal base, is the checksum.

Please note that in the NASA Prediction Bulletins the field assigned to the International Designator is usually empty.

The second line:
It is built according to the following rules:

Columns	Description
01-01	Line number of the elements (2)
03-07	Number of satellite
09-16	Inclination [degrees]
18-25	Right ascension of the ascending node [degrees]
27-33	Eccentricity (the decimal point is implied)

Columns	Description
35-42	Argument of perigee [degrees]
44-51	Mean anomaly [degrees]
53-63	Mean Motion [Revs per day]
64-68	Number of orbit to date
69-69	Checksum

The checksum is calculated using the same algorithm described above.

AMSAT format

This format is not fully standardised but generally the tracking programs interpret it correctly even if it comes from different sources.

The format is defined "user-friendly" as it is easily interpretable by people. Spaces are not significant and each element fills a separate line. The order is free, but each group must begin with the word "satellite" and a blank line indicates the end.

This format has no control characters but sometimes it is paired with its own checksum.

A typical example of AMSAT format is as follows:

Satellite:	AO-13
Catalog number:	19216
Epoch time:	94311.77313192
Element set:	994
Inclination:	57.6728 deg
RA of node:	221.5174 deg
Eccentricity:	0.7242728
Arg of perigee:	354.2960 deg
Mean anomaly:	0.7033 deg
Mean motion:	2.09727084 rev/day
Decay rate:	-5.78e-06 rev/day^2
Epoch rev:	4902
Checksum:	312

The checksum is computed with the same rules as the NASA-2 Line except that it is written in full and not just the last digit. Each character is considered. For example the "2" to "rev/day^2" enter into the calculation.

Glossary

Like any other field of human activity, the world of amateur satellite has its own specific language, often borrowed from relevant disciplines such as astronomy, physics and celestial mechanics. Below is a glossary of terms to help newcomers to the subject.

AMSAT
Stands for AMateur SATellites. This organisation brings together the various associations from around the world who share Amateur Satellites as a common interest. The first ever AMSAT was founded in the U.S. in 1969 with follow-up subsidiaries in many countries. Their websites contain a wealth of information on the subject of amateur satellites. Each website reflects the diversity and interests of the different countries

AOS
Stands for Acquisition Of Satellite. It indicates the time when the satellite rises above the horizon of an observer. See also: LOS

APOGEE
Is the most distant point of a satellite (or generic orbiting body) from the Earth, running a geocentric orbit (AKA maximum orbital distance). **Fig G.1** See also: perigee

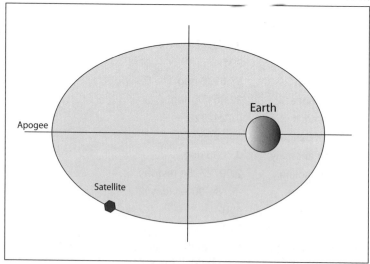

Fig G.1: Furthest point from the Earth along the orbit

ARISS
ARISS stands for Amateur Radio on the International Space Station. ARISS is a group of worldwide volunteers devoted to enhancing the use of amateur radio equipment installed aboard the International Space Station (ISS) for educational purposes. More specific information, current projects and ongoing activities can be found at the following links:

www.arrl.org/amateur-radio-on-the-international-space-station

www.ariss.net

www.ariss-eu.org

Columbus
Is a pressurised orbital module equipped and financed by the European Space Agency (ESA) as a space laboratory. Designed for an operational lifetime of 10 years, it is now docked with the ISS. This module

has been developed in collaboration with Alenia Spazio (Italy), now Thales Alenia Space (TAS), and EADS (Germany). Columbus has an almost cylindrical pressurised structure, with an outer diameter of about 4.5 meters and a length of about 8 meters, which allows for the housing of standard racks for scientific experiments in micro-gravity conditions. The original structure which dated back to the early 1980s included a self-orbiting laboratory in a "free flyer" configuration, meaning it was not coupled to any space infrastructure. The launch took place in 1992, the fifth centenary of the discovery of America, hence the name Columbus inspired by Christopher Columbus. The evolution of this project, with all its ups and downs as well as delays, allowed Columbus to become an integral part of ISS in 2008.

There is amateur radio equipment and antennas installed in Columbus. See also: ARISS

Down-link

Down-link is the radio link between the satellite and ground station (direction space-to-ground). See also: up-link and link budget

Eccentricity (of the orbit)

The parameter indicating how much the orbit deviates from a perfect circle. The orbital eccentricity (e) is strictly defined for all circular orbits, such as elliptic, parabolic, or hyperbolic as follows:

for circular orbits:	e = 0
for elliptic orbits:	0 <e <1
for parabolic paths:	e = 1
for hyperbolic trajectories:	e> 1

Ephemeris

Ephemeris (taken from the Greek ephemeros which means daily) are tables that contain calculated values for a particular time interval of different astronomical variables such as magnitude, orbital parameters, coordinates of planets, comets, asteroids, artificial satellites, magnitude of variable stars, etc.

For the radio amateur, ephemeris means a list of data for satellite tracking.

Below (**Fig 1.2**) is an example of data for a typical orbiting body like ISS as calculated from my home town:

AMSAT Online Satellite Pass Predictions - ISS								
View the current location of ISS								
Date (UTC)	AOS (UTC)	Duration	AOS Azimuth	Maximum Elevation	Max El Azimuth	LOS Azimuth	LOS (UTC)	
13 Oct 13	17:30:13	00:11:00	246	68	337	60	17:41:13
13 Oct 13	19:07:37	00:10:19	279	25	335	65	19:17:56
13 Oct 13	20:44:58	00:10:33	297	29	29	88	20:55:31
13 Oct 13	22:21:50	00:10:59	299	80	219	124	22:32:49
13 Oct 13	23:59:04	00:09:00	285	12	225	172	00:08:04
14 Oct 13	15:06:54	00:08:56	187	12	127	76	15:15:50
14 Oct 13	16:42:12	00:10:52	235	79	151	61	16:53:04
14 Oct 13	18:19:23	00:10:33	272	29	4	63	18:29:56
14 Oct 13	19:56:51	00:10:19	294	25	350	80	20:07:10
14 Oct 13	21:33:48	00:10:54	300	67	30	114	21:44:42

Fig G.2: Example of ephemeris. Credits: www.amsat.org

Hamsat - Amateur Radio Satellites Explained

EnviSat (declared 'dead' by ESA May 2012)

Is a satellite (the name stands for ENVIronmental SATellite) developed by ESA to study the Earth's environment. Launched on 1 March 2002, it was placed in a sun-synchronous polar orbit at an altitude of 790km (± 10km). Its orbital period is 101 minutes and it performs a complete cycle in 35 days. It is about the size of a large bus, 25 × 7 × 10 meters with a mass exceeding 10 tonnes. Of particular relevance among the results obtained, in March 2007, after three years of analysis of the Sciamachy instrument, that scientists discovered that the mathematical models of methane production values, were significantly underestimated. The satellite data was then used to adjust the models in order to reduce the error.

ESA

Stands for European Space Agency. As the EU's gateway to space, ESA organises and coordinates all the activities related to the EU's space study and exploitation. ESA is composed of eighteen member countries which are: Austria, Belgium, Denmark, Finland, France, Germany, Greece, Ireland, Italy, Luxembourg, Norway, the Netherlands, Portugal, United Kingdom, Czech Republic, Spain, Sweden and Switzerland. Poland, Romania and Hungary cooperate with ESA on specific projects as well as Canada.

The ESA headquarter is located in Paris, with the following operating branches:

- **ESTEC**: the European Space Research and Technology Centre is ESA's centre of spacecraft projects and technological development based in Noordwijk, the Netherlands.
- **ESOC**: the European Space Operations Centre. Located near Darmstad, Germany, ESOC is the control centre for ESA satellites.
- **EAC**: the European Astronauts Centre. Located in Cologne, Germany, EAC is the training centre of cosmonauts for future missions.
- **ESRIN**: the European Space Research Institute. Located in Frascati, Italy, ESRIN handles the collection, storage and distribution of Earth satellite observation data gathered by ESA and its partners.
- **ESAC**: the European Space Astronomy Centre. Located in Villafranca, Spain, ESAC is responsible for the collection, storage and retrieval of astronomical and planetary mission data.

Ground Station

Is a generic transceiver (or even just receiving) station located on the Earth's surface capable of supporting satellite traffic.

IHU

Acronym for Integrated Housekeeping Unit, an IHU is the satellite's computer centre responsible for the management of all the resources on board.

Inclination (of the orbit)

Each closed orbit lies in a plane known as the orbital plane, which passes through the centre of our planet. The angle between this plane and the ecliptic indicates the inclination of the orbit (**Fig G.3**) with respect to the Earth.

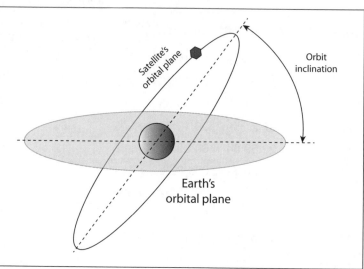

Fig G.3: Inclination of the orbit

Glossary

By definition, it can vary from 0 degrees (when the satellite orbit is around the equator) to 180 degrees (specular situation). The orbits with an inclination close to 90 degrees are called polar, as the satellite flies over just the two poles of the earth.

ISS

ISS stands for International Space Station. It represents a permanent outpost of human presence in space as it has been continuously inhabited since 2 November, 2000 by at least two astronauts. Crew change takes place at least every six months. It can be considered the evolution of MIR and Skylab programs as a joint project of five space agencies. The space station is a modular vehicle built and developed over the last 15 years with the help of the various agencies involved whose development has been interrupted several times due to tragic accidents and repeated technological and economic difficulties. In June 2010, the space station was composed of the following modules and elements:

Element	Flight	Carrier	Lunch date	Mass kg
Zarja	1 A/R	Proton	20 November 1998	19 323
Unity	2A - STS-88	Endeavour	4 December 1998	11 612
Zvezda	1R	Proton	12 July 2000	19 050
Z1 Truss	3A - STS-92	Discovery	11 October 2000	8 755
P6 Truss - solar panels	4A - STS-97	Endeavour	30 November 2000	7 700
Destiny	5A - STS-98	Atlantis	7 February 2001	14 515
Canadarm2	6A - STS-100	Endeavour	19 April 2001	4 899
Joint Airlock	7A - STS-104	Atlantis	12 July 2001	6 064
Docking Compartment - Pirs Airlock	4R	Soyuz	14 September 2001	3 900
S0 Truss	8A - STS-110	Atlantis	8 April 2002	13 971
Mobile Base System	UF-2 - STS-111	Endeavour	5 July 2002	1 450
S1 Truss	9A - STS-112	Atlantis	7 October 2002	14 124
P1 Truss	11A - STS-113	Endeavour	24 November 2002	14 003
External Stowage Platform (ESP-2)	LF 1 - STS-114	Discovery	26 July 2005	2 676
P3/P4 Truss – Solar panels	12A - STS-115	Atlantis	9 September 2006	16 183
P5 Truss	12A.1 - STS-116	Discovery	10 December 2006	1 864
S3/S4 Truss – Solar Panels	13A - STS-117	Atlantis	8 June 2007	16 183
External Stowage Platform (ESP-3)	13A.1 - STS-118	Endeavour	18 August 2007	2 676
S5 Truss	13A.1 - STS-118	Endeavour	8 August 2007	1 864
Harmony Node 2	10A - STS-120	Discovery	24 October 2007	14 288
Columbus	1E - STS-122	Atlantis	7 February 2008	10 300/19 300
Japanese Experiment Module - ELM PS	1J/A - STS-123	Endeavour	11 March 2008	8 386
Japanese Logistic Module - JLM-PM	1J - STS-124	Discovery	31 May 2008	14 800
S6 Truss – Solar Panels	15A - STS-119	Discovery	20 March 2009	15 900
Japanese Experiment Module - ELM-EF	2J/A - STS-127	Endeavour	15 July 2009	4 000
Poisk - MRM2	5R	Soyuz	10 November 2009	3 670
Node 3 e Cupola	20A - STS-130	Endeavour	8 February 2010	
Mini-Research Module 1	39A - STS-132	Atlantis	14 May 2010	5 075

LEO

Acronym for Low Earth Orbit, *LEO* is a nearly circular orbit with an altitude equal to what is between the outer layer of the atmosphere and the Van Allen belts. It is the typical orbit for space missions such as the Space Shuttle or the International Space Station as well as the majority of amateur satellites. A body that orbits in a *LEO* has a revolution period of around 90 minutes and a speed of around 27,000km/h.

By extension, *LEO* also indicates the name of the satellite telecommunication system in which the satellites are placed on low orbits (200-2000 miles). The lower limit of 200km altitude is imposed by the atmosphere drag: at lower altitude the friction with atmosphere would produce a rapid orbital decay and consequent destructive satellite re-entry into the atmosphere. The upper limit of about 2000km is instead imposed by the Van Allen belts which, if crossed repeatedly, would cause a strong radiation exposure for the satellite that could affect the proper operation of the equipment on board. The main advantage of these systems is the limited propagation delay time (20-25ms) of the communications link and the relatively low cost of launching. For the latter reason *LEO* is the kind of orbit now universally chosen for amateur satellites, where the budget is clearly limited.

Link budget
Is the equation that takes into account the parameters of the radio link to or from the satellite and it is used to estimate the available connection quality. Usually, we take into consideration the transmitted power, the gain of the antennas, the distance of the satellite from the ground station, the type of modulation, the bandwidth, the atmosphere attenuation, and so on. The result is a number (usually in dB) that expresses the ability to perform the radio link in those conditions. See also: down-link and up-link.

LOS
Acronym for *Loss Of the Satellite*. *LOS* indicates the time when the satellite passes below the observer's horizon. *See also*: AOS

MIR
MIR is the transcript of 'Мир', the Russian word meaning "peace" or "world". Formerly Soviet then Russian, MIR used to be an important space station, active between 1986 and 2001. The record for the most time spent in space is held by the cosmonaut Valeri Polyakov who spent about 437 days and 18 hours in MIR. MIR recorded 9 years and 257 days of uninterrupted human presence on board, making this another record. see Fig G.4

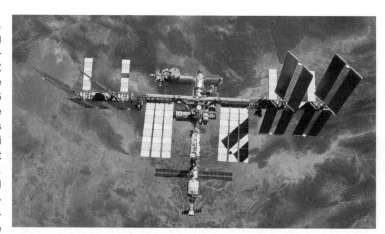

Fig G.4: The Soviet/Russian space station MIR. Courtesy: NASA

Molniya (orbit)
Meaning "flash" in Russian, Molniya is a special type of orbit designed and used by military communications satellites of the former Soviet Union. It is an elliptical orbit highly eccentric with an inclination of about 63.4 degrees and an orbital period of 12 hours. Its name comes from the earliest Molniya Soviet satellites that were placed on this kind of orbit during the 1960s: this was a constellation of three satellites capable of covering the entire territory of the Soviet Union.

Orbit
Orbit is the trajectory of a celestial body, an artificial satellite or a spacecraft in space. It can be divided into several categories depending on the scope and application considered. The main subdivisions are:

Open / closed
- Based on the energy possessed by the body, the orbits can be closed and periodic or open and non-periodic.

Glossary

- Elliptical trajectory: the orbit is closed and is an ellipse if the specific orbital energy of the body is less than zero. All the planets of the solar system and all their satellites have elliptical orbits. The circular orbit is a special case of an elliptical orbit.
- Hyperbolic trajectory: the orbit is open and is a hyperbola if the specific orbital energy of the body is greater than zero. The space probes sent towards the outside of the solar system and the portions of orbits of the probes sent to the outer planets (such as Galileo and Cassini spacecraft) while using the slingshot effect of the internal planets, have hyperbolic orbits.
- Parabolic trajectory: it represents the separation element between the family of closed orbits and open orbits without global energy for its own bodies.

According to the inclination:
- Equatorial orbit: when the inclination is approximately zero (i.e. the geostationary orbit).
- Polar orbit: the inclination is nearly 90 degrees. The polar-orbiting satellites have the characteristics of being able to fly over the whole globe.
- Ecliptic orbit: if the inclination of the orbit coincides with the ecliptic of the planet.
- Retrograde orbit: when the inclination is greater than 90 degrees.

According to the use:
- Commercial orbit
- Parking orbit
- Molniya orbit
- Sun-synchronous orbit
- Transfer orbit

According to the altitude:
- Low Earth Orbit: in which is located for instance the International Space Station
- Medium Earth Orbit: used by the satellites for navigation systems (GLONASS, Galileo and GPS).
- Highly Elliptical Orbit: used sometimes by amateur satellites intended to deliver worldwide communications.

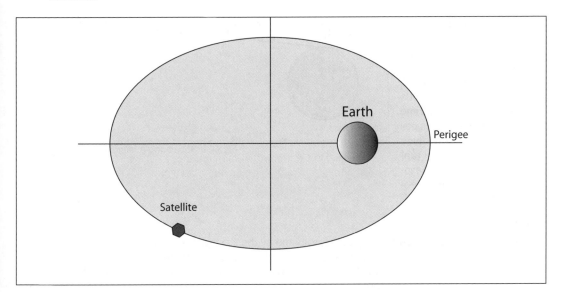

Fig G.5: Closest point from the Earth along the orbit

Hamsat - Amateur Radio Satellites Explained

- Geostationary orbit: at an altitude of 36000km (known as "Clarke Belt") the satellites can remain apparently in a fixed position related to the earth's surface. Satellites for telecommunications and radio and television broadcasting (TV-sat) lie in this orbit.

OSCAR
It stands for "Orbiting Satellite Carrying Amateur Radio". Some amateur satellites which meet specific requirements by AMSAT are named with this designation.

Perigee
Perigee is the closest point to the earth of a geocentric orbit (minimum orbital distance) of a satellite (**Fig 1.5**). *See also:* Apogee

Satellite (artificial)
Artificial (man-made) object intentionally placed in orbit around a celestial body.

Spin
Rotational motion of the satellite, usually intended to stabilise its orientation towards the ground

Squint angle
Is the angle that describes "how much" the satellite antennas are pointed towards the ground station. The optimal value of 0 degrees indicates that the antenna is pointing perfectly at our station, creating the best conditions for the connection. The value of 90 degrees indicates that the antenna "shows the side" to the ground station and 180 degrees while the antenna is facing into outer space (**Fig G.6**).

Up-link
Generally, it is the link between the ground station and satellite (earth-to-space direction).

See also: down-link, link budget

UTC
Stands for Universal Time Coordinated. UTC is the reference time zone based on which all the other time zones are calculated all over the world. UTC derives from Greenwich Mean Time (GMT) and sometimes is still called GMT. This new name came about for 'political reasons' and is no longer associated with a specific geographic location.

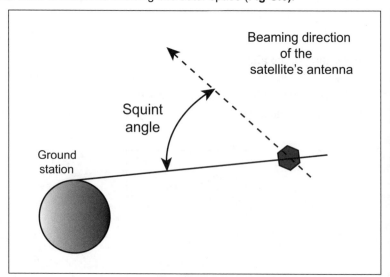

Fig G.6: Squint angle: alignment of the antennas

Van Allen (Belts)
This is a Torus of charged particles (plasma) retained by the Earth's magnetic field at high altitude. Their existence was theorised long before the space age and achieved experimental confirmation from the results of the missions of Explorer 1 (31 January 1958) and Explorer 3, under the supervision of Professor James Van Allen (hence the name).

The Van Allen Belts actually consist of two distinct areas surrounding our planet, the innermost belt mainly consisting of protons while the outer belt consists of free electrons. It is commonly believed that the Van Allen Belts are the result of the collision of the solar wind with the earth's magnetic field.

The Earth's atmosphere limits the extension of the lower boundary at an altitude of 200-1000km while its upper boundary is however rather vague. It does not pass over the 40,000km from the earth's surface. The bands are located in an area that covers approximately 65 degrees north and south of the celestial equator.

They represent a serious danger for crossing spacecraft and in particular are very aggressive for a multiplicity of electronic devices such as photovoltaic panels, integrated circuits and sensors. Satellites whose orbit is expected to be within the Van Allen Belts are designed and built with very robust and expensive protection systems.

Zenith

Zenith is the point on the celestial sphere that is directly above the observer. The diametrically opposite point is called Nadir.

This term derives from the Arabic "samt al-ra's" or "samt al-ru'ūs" ("head direction" or "heads"), where "samt" is a calque of the Greek "semeion". Plato Tiburtinus translated this into Latin - "zenith capitis" or "zenith capitum."

Appendix 1
Associations Dedicated To Amateur Radio Satellites

Since the era of the OSCAR Project, the earliest founder of a team of amateur satellites and space communications enthusiasts, many other groups and organisations have arisen in various parts of the globe with the common goal of bringing together enthusiasts to develop new projects independently as well as in collaboration with other associations.

Logo of Project OSCAR team, the very first group of enthusiasts of amateur satellite communications

Undoubtedly, the most popular, prestigious and authoritative association is **AMSAT**.

AMSAT stands for AMateur SATellites. Under this name various enthusiast associations all over the world are gathered together. The first one was founded in the US (AMSAT-NA) then several subsidiaries in other countries followed over the years.

Their websites are a mine of information on the subject of amateur satellites. The brand is recurrent and distinctive with local customisation. For example, following is a list of the most interesting and historically important ones:

Logo	Country	Website
	USA	www.amsat.org
	United Kingdom	www.amsat-uk.org

Appendix 1

Logo	Country	Website
	Germany	www.amsat-dl.org
	Sweden	www.amsat.se
	Finland	www.rats.fi/suomi/amsat-oh/amsat-oh.html
	Australia	www.amsat-vk.org
	New Zealand	www.amsat-zl.org.nz

Logo	Country	Website
	South Africa	www.amsatsa.org.za
	India	www.amsatindia.org
	Japan	www.jamsat.or.jp
	Venezuela	www.amsat-yv.org
	Argentina	www.amsat.org.ar

Bibliography

Atlante geografico, *del Dizionario Enciclopedico Moderno*, Milano, Edizioni, Labor, 1943

Charles W. Bostian et al. *New markets and policy considerations*, www.wtec.org/loyola/satcom/c4_s1.htm

Davidoff, Martin, *The Satellite Experimenter's Handbook*, Newington, CT: ARRL, 1984

Doug McArthur, VK3UM, *Sunnoise measurement*, GippsTech, 2008

Dragoslav Dobridic, YU1AW, *Determining the Parameters of a Receive System in Conjunction with Cosmic Radio Sources*, VHF Communications -1/1984

Frank Sperber, DL6DBN, *AO-40 Mini-Glossary*, www.amsat-dl.org/journal/adlj40ge.htm

Gould Smith WA4SXM, *AO-51 Development, Operation and Specifications*, AMSAT, 2004

Hans J. Hartfuss DL2MDQ, *Radio Astronomy for the VHF/UHF Radio Amateur*, VHF Communications, 1988

Heinz Hildebrand, *Das Mode-J-Filter*, 1986

Herbert Taub, Donald Schilling, *Principles of communication systems*, 2nd edition, McGraw-Hill International Editions, 1986

Hewes & Jessop, *Radio Data Reference Book*, 5th edition, RSGB, 1985

International Telecommunication Union, *Recommendation ITU-R P.372-8: Radio noise*, 2003

James Miller, *A PSK Telemetry Demodulator for OSCAR 10*, Ham Radio, Apr 1985, pp. 50-51, 53-55, 57-62

James Miller, *Measure AO-13 Squint Directly*, The AMSAT Journal, Vol. 16, No. 1, Jan/Feb 1993, p. 20

James Miller, *Managing Oscar-13*, The AMSAT Journal, Vol. 17, No. 1, Jan/Feb 1994

James Miller, *The Re-Entry of Oscar-13*, The AMSAT Journal, Vol. 18, No. 3, May/Jun 1995

Jan King, et al. *OSCAR at 25: The Amateur Space Program Comes of Age*, QST, Dec 1986

Jansson, Richard, *Spacecraft Technology Trends in the Amateur Satellite Service*, Ogden, UT: Proceedings of the 1st Annual USU Conference on Small Satellites, 1987

John A. Magliacane, *Spotlight On: AMSAT-OSCAR-13*, The AMSAT Journal, Vol. 15, No. 2, Mar/Apr 1992, p. 17

John Branegan, *Space Radio Handbook*, RSGB, 1991

Karl Meinzer, *The Orbital Decay of AMSAT-OSCAR-13?*, The AMSAT Journal, Vol. 13, No. 3, Jul 1990, p. 1

Keith Baker and Dick Jansson, *Space Satellites from the World's Garage -- The Story of AMSAT*, atti del convegno National Aerospace and Electronics Conference (Dayton, Ohio, 23-27 May 1994)

Keith Berglund, *Beginner's Guide to AO-13*, AMSAT, 1989

Mark Wade, *OSCAR*, www.astronautix.com/craft/oscar.htm

Martin Davidoff, *Satellite Experimenter's Handbook*, 2nd edition, Newington, CT, ARRL, 1990

Martin Davidoff K2UBC, *The Radio Amateur's Satellite Handbook Revised,* 1st edition, Newington, CT, ARRL, 2001

G/T Maximisation of a Paraboloidal Reflector Fed by a Dipole-Disk Antenna with Ring by Using the Multiple-Reflection Approach and the Moment Method, IEEE Transactions on Antennas and Propagation, vol. 45. n. 7

Peter Gulzow DB2OS, *AMSAT OSCAR-13 Telemetry Block Format*, OSCAR News, No. 73, Oct 1988, pp 8-14

Pierluigi Poggi, *Rock solid preamplificatore 430MHz*, Radio Kit Elettronica, April 2009

Stratis Caramanolis, *OSCAR*, RSGB, 1976

Syed Idris Syed Hassan, *Satellite Link Impairment due to Rain and Other Climatic Factors*

Various authors and contributors
JAS-1 Satellite Handbook, JARL, 1985

AMSAT Satellite Report, No. 177, 178, 179, Jun 8, Jul 5, Jul 18, 1988

Phase 3C System Specifications, Part 1 and Part 2, Amateur Satellite Report, no. 177, 8 Jun 1988, pp. 3-4

Satellite Anthology, 2nd edition, Newington, CT, ARRL, 1992

Proceedings of the AMSAT-North America 2006 Space Symposium, AMSAT, 2006

Per-Simon Kildal, *Svein A. Skyttemyr, and Ahmed A. Kishk*, 1997

en.wikipedia.org/wiki/AMSAT

en.wikipedia.org/wiki/Mir

www.amsat.org

www.esa.int

www.funcubedongle.com

www.spacetoday.org/Satellites/Hamsats/Hamsats1960s.html, 2006

Efficiency and Temperatures, http://www.med.ira.inaf.it/ManualeMedicina/English/5.%20Efficiency.htm

Bibliography

Italian websites

Francesco Grappi, *Satellite AO-40 (Phase 3D)* www.arimodena.it/Members/iw4dvz/il-mitico-indimenti-cabile-ao-40

www.asi.it

it.wikipedia.org/wiki/Apsidi

it.wikipedia.org/wiki/Cifra_di_rumore

it.wikipedia.org/wiki/Columbus_orbital_facility

spaceradioandmore.blogspot.com/

www.br73.net/sputnik.htm

www.rs-components.it

Propagazione delle onde radio, www.air-radio.it/propagazione.html

Index

A

Amateur Radio Satellites	112
AMSAT	104
	112
Antenna Rotator	31
Antennas	82
AO-51	7
AOS	104
APOGEE	104
Argument of Perigee	98
Ariane	
Ariane 4	15
Ariane 5	15
ARISS	104
Ascension of a Satellite	98
Associations	112
Atmosphere	19
Atmospheric absorption	23
Attitude	100
Aurora Borealis	24
Azimuth	48

B

Bahn Coordinates	100
Bibliography	115

C

Cables	44
Columbus	104
Connectors	44
Contents	iii
Converter	46
CubeSat	7

D

Demodulation	45
D layer	22
Doppler Effect	85, 87, 89
Down-link	105
Drag	100

E

Eccentricity	99
E layer	22
Elements	95, 97, 99, 101, 103
Ephemeris	105
Epoch	97
ERP	27
ESA	106
Exosphere	22

F

Faraday rotation	25
Field aligned irregularity	25
F layer	22
Foreword	iv, v
Frequency	7
Front-end bandwidth	35
FunCube	46

G

Gain	34
Glossary	104

Ground-Space-Ground Propagation   17, 19, 21, 23, 25
Ground Station 31, 33, 35, 37, 39, 41, 43, 45, 47, 49, 51, 53, 55, 57, 106

H

HAMSAT	7
Heterosphere	19
HO68 Hope OSCAR	7
Homosphere	19

I

IARU Band Plan	93
IHU	106
Inclination	106
Inverting	87
Ionisation	22
Ionospheric reflection	25
ISS	107

K

Keplerian Elements	95, 97, 99, 101, 103
Kepler's First Law	95
Kepler's Second Law	96
Kepler's Third Law	96

L

Launch Bases	14
LEO	107
Link budget	27, 29, 108
LOS	108

M

Main Orbital Carriers	15
Mean Anomaly	100
Mean Motion	99
Mesosphere	21
Meteor Scatter	24
MIR	108
Modes	93
Molniya	108

N

Noise Figure	34
non-inverting	87

O

Orbit	108
Orbital Inclination	97
OSCAR	3, 110
AO73 FunCube	7
HO68 Hope OSCAR	7
OSCAR 5	5
OSCAR 6	5
OSCAR 7	5
OSCAR 8	5
OSCAR 9	6
OSCAR 10	6, 101
OSCAR 12	6
OSCAR 13	6
OSCAR 40	7
OSCAR I	4
OSCAR II	4
OSCAR III	4
OSCAR IV	5

P

Path	28
Perigee	98
Polarisation	73
Preamps	34
Propagation	17, 19, 21, 23, 25

R

Rain scattering	23
Receivers	44
refraction	23
Rotator	31

S

Satellite Phase	3
Satellites Modes	93
SDR	46

Index

Selectivity	45	Troposphere	20
Spin	110	Tropospheric scattering	23
Sporadic E	25		
Sputnik	5	**U**	
RS-3, RS-4, RS-5, RS-6, RS-7 and RS-8	6	UoSAT	6
Sputnik 17	6	UoSAT-OSCAR 9 (UoSAT 1)	6
Squint angle	110	Up-link	110
Stratosphere	21	UTC	110

T

V

The layers of the atmosphere	19	Van Allen	110
Thermosphere	21		
Tracking system	48	**Z**	
Transceiver	47	Zenith	111
Transequatorial Anomaly	25		

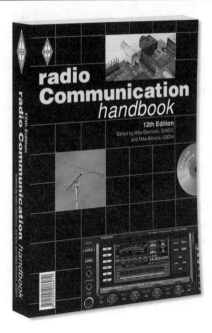

RSGB Radio Communication Handbook

Whether you are an operator who wants to know what goes on 'under the hood', an avid constructor or someone keen to go on learning about electronics and communication radio, the *RSGB Radio Communication Handbook* is the one book you need. With nearly 600,000 words plus 2,000 illustrations and tables crammed into 864 pages.

This edition has been thoroughly updated with many expanded features. You will find a completely revised HF Receivers chapter and the Propagation chapter has also been completely rewritten with an eye on improving clarity and understanding. There are new authors for the sections on Practical Microwave Antennas and the Low Frequencies. Based on the popular Homebrew column in *RadCom* there is a new chapter that follows the design and construction of an HF transceiver, and provides a host of valuable information and circuit ideas. Expanded chapters include Practical VHF-UHF Antennas, The Great Outdoors, Morse and Digital Communications. Many other chapters have had new, revised and updated parts. Since the last edition, there has been more use of microcomputers, such as the Raspberry Pi, in amateur radio projects; a new amateur band at 472kHz has also been created. These changes have been incorporated into the relevant chapters. Peter Hart's comparison chart of HF receiver performance has been updated, there's more on optical communications and new datamodes have been included.

FREE CD
Enclosed with the book there is a fully searchable electronic version of the Handbook in PDF format. You will find many bonus chapters, samples of other RSGB publications and even a host of useful amateur radio software.

Size: 210x297mm (A4), 864 pages
ISBN: 9781 9050 8697 9
Non Members' Price: £32.99
RSGB Members' Price: £27.99

E&OE All prices shown plus p&p

Radio Society of Great Britain www.rsgbshop.org
3 Abbey Court, Priory Business Park, Bedford, MK44 3WH. Tel: 01234 832 700 Fax: 01234 831 496